Experiment with the
Sense HAT, 2nd Edition

Raspberry Pi Essentials: *Experiment with the Sense HAT*, 2nd Edition
by Raspberry Pi Foundation Learning Team
ISBN: 978-1-916868-40-3
Copyright © 2025 Raspberry Pi Ltd
Published by Raspberry Pi Ltd, 194 Science Park, Cambridge, CB4 0AB
Raspberry Pi Ireland Ltd, 3 Dublin Landings, D01 C4E0, compliance@raspberrypi.com

Editor: Lucy Hattersley
Copy Editor: Sarah Cunningham
Interior Designer: Sara Parodi
Production: Brian Jepson
Photographer: Brian O'Halloran
Illustrator: Sam Alder
Graphics Editor: Natalie Turner
Publishing Director: Brian Jepson
Head of Design: Jack Willis
CEO: Eben Upton

April 2025: Second Edition
January 2021: First Edition

The publisher and contributors accept no responsibility in respect of any omissions or errors relating to goods, products, or services referred to or advertised in this book. Except where otherwise noted, the content of this book is licensed under a Creative Commons Attribution-NonCommercial-ShareAlike 3.0 Unported (CC BY-NC-SA 3.0).

Table of Contents

v   Welcome to Experiment with the Sense HAT
vii   Colophon

Chapter 1
1   What is the Sense HAT?
The special add-on board that lets your Raspberry Pi interact more with the world around it — as seen on the ISS!

Chapter 2
7   High-flyers
British ESA astronaut Tim Peake wasn't the only Brit aboard the International Space Station in 2015. Let's look at how Raspberry Pi first got into the sky and what it means for future generations.

Chapter 3
19   Get started with the Sense HAT
Now that you know all about the Sense HAT and the part it plays in Astro Pi, it's time to learn how you can actually use one for your own stellar projects

Chapter 4
37   Random sparkles
Create amazing random sparkles with your Sense HAT as you learn how to set random pixels in various colours on its LED display

Chapter 5
45   Make a digital twist on the Magic 8 Ball
Bring the time-honoured tradition of shaken-not-stirred fortunes to the Sense HAT, making use of its built-in motion sensors

Chapter 6
51 Interactive pixel pet
Create a pint-sized pocket monster living in the digital world of a Raspberry Pi Sense HAT

Chapter 7
61 Sense HAT data logger
Use your Sense HAT to take environmental readings, so you can walk around the house saying "sensors indicate…"

Chapter 8
73 Flappy Astronaut
Create your own clone of the Flappy Bird game using your Raspberry Pi, a Sense HAT, and some Python code

Welcome to *Experiment with the Sense HAT*

Space exploration is fascinating and inspiring for children and adults alike. With the tiny Raspberry Pi computer helping to change the world little by little, it was only a matter of time before it went into space to help out there as well. The Raspberry Pi Sense HAT aids two Raspberry Pi computers on board the International Space Station in their extraterrestrial mission: it sits on top of each Raspberry Pi, equipped with lights and sensors that allow the Raspberry Pi to interact with the world around it.

In this book, we help you figure out exactly what the Sense HAT is and how you can use it to make your projects and dreams a reality. It's an incredibly versatile and flexible bit of kit with plenty of obvious uses, along with a huge number of less obvious ones, that you'll love to make and share. Updated for the latest Raspberry Pi devices and hardware, this book has everything you need to get started.

You can find example code and other information about this book, including errata, in its GitHub repository at **rpimag.co/sensebookgit**. If you've found what you believe is a mistake or error in the book, please let us know by using our errata submission form at **rpimag.co/sensebookfeedback**.

Colophon

Raspberry Pi is an affordable way to do something useful, or to do something fun.

Democratising technology — providing access to tools — has been our motivation since the Raspberry Pi project began. By driving down the cost of general-purpose computing to below $5, we've opened up the ability for anybody to use computers in projects that used to require prohibitive amounts of capital. Today, with barriers to entry being removed, we see Raspberry Pi computers being used everywhere, from interactive museum exhibits and schools to national postal sorting offices and government call centres. Kitchen table businesses all over the world have been able to scale and find success in a way that just wasn't possible in a world where integrating technology meant spending large sums on laptops and PCs.

Raspberry Pi removes the high entry cost to computing for people across all demographics: while children can benefit from a computing education that previously wasn't open to them, so too can the many adults who have historically been priced out of using computers for enterprise, entertainment, and creativity. Raspberry Pi eliminates those barriers.

Raspberry Pi Press

store.rpipress.cc

Raspberry Pi Press is your essential bookshelf for all things computing, gaming, and hands-on making. As the official publishing imprint of Raspberry Pi Ltd, we publish a range of titles to help you make the most of your Raspberry Pi hardware. Whether you're building a PC or building a cabinet, you can discover your passion, learn new skills, and make awesome stuff with our extensive collection of books and magazines.

Raspberry Pi Official Magazine

magazine.raspberrypi.com

Raspberry Pi Official Magazine is a monthly publication for makers, engineers, and enthusiasts who love to create with electronics and computer technology. Each issue is packed with Raspberry Pi-themed projects, tutorials, how-to guides, and the latest community news and events.

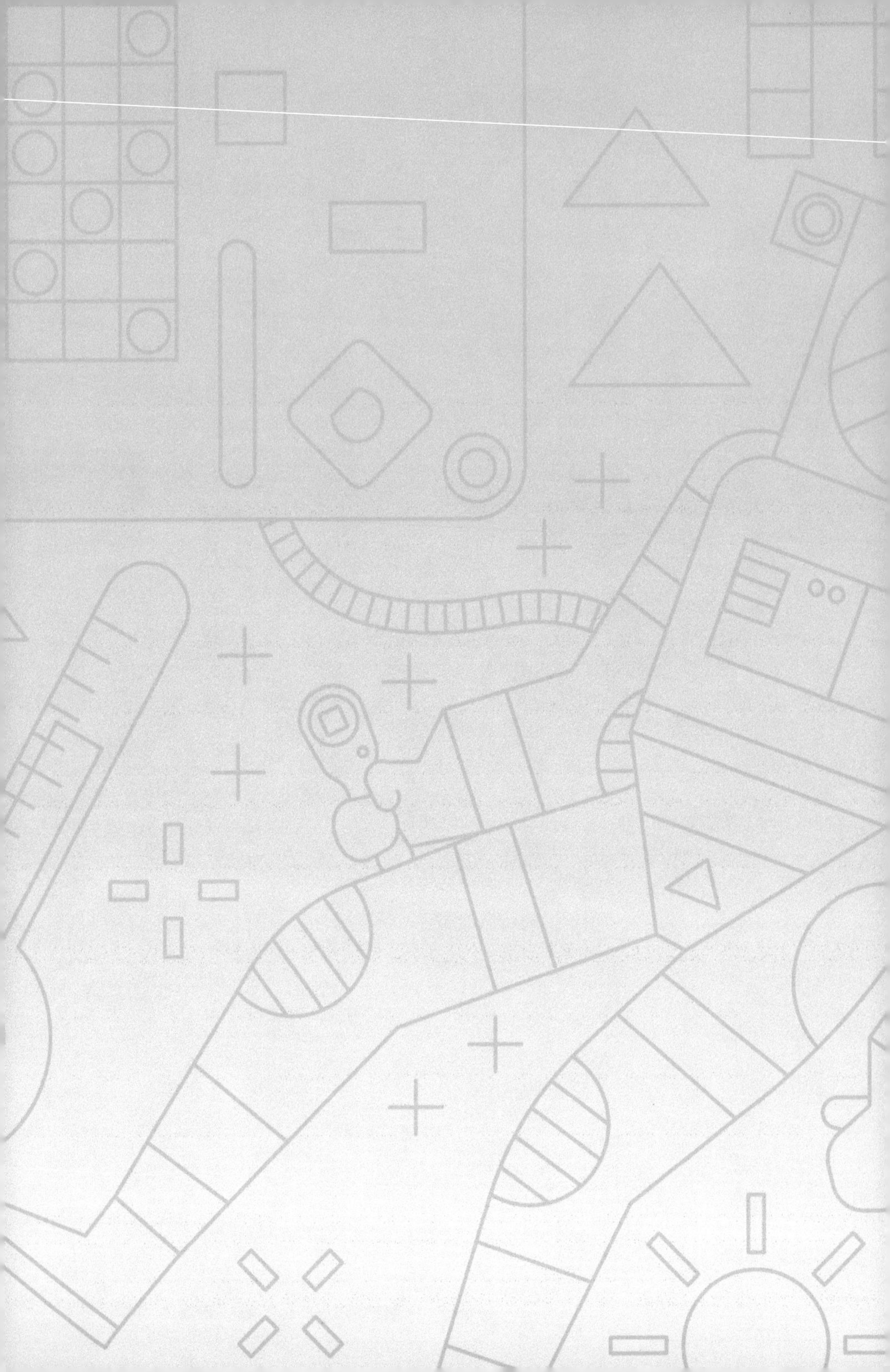

Chapter 1
What is the Sense HAT?

The special add-on board that lets your Raspberry Pi interact more with the world around it — as seen on the ISS!

A Raspberry Pi computer can do many things, thanks to its size, portability, and ability to connect to the internet; you can even connect a Raspberry Pi to other electronic devices via the pins on its GPIO header, enabling it to interact with the world around you. One such device is the Sense HAT V2 (**Figure 1-1**), which packages a collection of sensors and output devices into one device, making it one of the easiest ways to get started with *physical computing*.

As the name suggests, physical computing is all about controlling things in the real world with your computer programs, using hardware alongside software. When you set the program on your washing machine, change the temperature on your programmable thermostat, or press a button at the traffic lights to cross the road safely, you're using physical computing.

At the top edge of a Raspberry Pi circuit board, or the back of a Raspberry Pi 400 or 500, you'll find two rows of metal pins known as the GPIO (general-purpose input/output) header. This header allows you to connect hardware, such as LEDs and switches, to the Raspberry Pi and control them using programs you create. The pins can be used for both input and output, and they can also communicate with add-on hardware, including the Sense HAT.

Figure 1-1   The Sense HAT V2

The Sense HAT is a sophisticated add-on board for the Raspberry Pi. While HAT is an acronym (*Hardware Attached on Top*), it does act, in a way, like a hat for your Raspberry Pi. The Sense HAT contains a suite of sensors that allows the Raspberry Pi to detect the world around it, along with an array of LEDs on top that can be used to display information about what the board can sense. It also has a little joystick.

The Sense HAT is a vital component of Astro Pi, the specially adapted educational Raspberry Pi computer sent up to the International Space Station with British ESA astronaut Tim Peake (**Figure 1-2**) in 2015 to run code created by children. This wasn't what the HAT was originally designed for, though, as the Sense HAT's Project Lead Jonathan Bell explains:

"I sort of hijacked a pet project of James [Adams'] and turned it into a space-faring board," says Jonathan. James Adams is the Chief Operating Officer and Hardware Lead at Raspberry Pi, and along with Jonathan, was one of the main driving forces behind the Sense HAT.

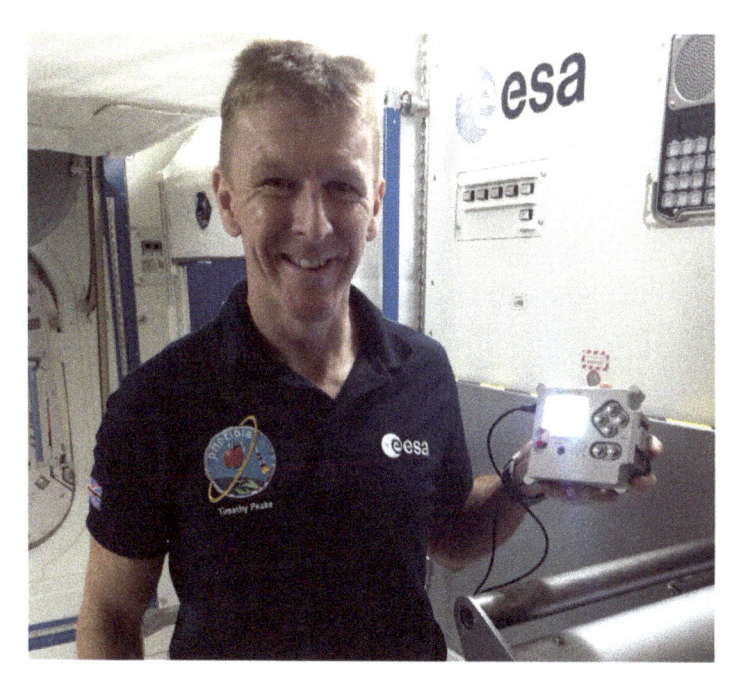

Figure 1-2 British ESA astronaut Tim Peake. Photo: ESA/NASA

"Effectively, we wanted to produce a board that would be a neat, fun example of how to design a HAT," Jonathan continued. "It was an exercise in how to design a HAT which could be put into mass production: how would somebody go about doing that so hundreds of thousands of HATs could be made, and how would we design the board to deal with that?"

Halfway through development, what was once a relatively basic HAT had some sensors added onto it, much like the kind used in mobile phones. "Eventually we said, 'Hang on a minute, what happens if we put loads of sensors on this thing and turn it into a kind of cool toy?'."

When the Sense HAT was eventually completed, it had three key sensors: separate pressure and humidity sensors that can also measure temperature, and a motion sensor that contains an accelerometer, a gyroscope, and a magnetometer. These sensors are joined on the board by the 8×8 LED screen and the joystick.

Sense HAT V2 (**Figure 1-3**) was released in 2022, adding a TCS3400 colour (red, green, blue, clear, and infrared) ambient light sensor; V2 also disables the LEDs if a 3.3V voltage is not present.

Each of the sensors, the LED screen, and the joystick can all work independently of each other, as well as all together at once. You could have the Sense HAT keep track of the temperature throughout the day or simply make the LED screen display little images for you; it's very flexible to use!

All of this is accessible on a Raspberry Pi just by popping the Sense HAT onto the GPIO pins and using the right Python code, which is what the Astro Pi units on the ISS are doing.

"The Astro Pi experiments make good use of the HAT itself," Jonathan told us. "Some of them in quite unusual ways. We have a few Easter eggs up there, which you'll have to find out about, but there have been some ingenious uses of the sensors. One of the experiments that caught our eye in terms of sensing was one that attempted to detect an astronaut. The astronaut detector sits there monitoring the humidity, and if there is a certain percentage change in humidity in the module, it determines there's an astronaut present. It flashes a message on the LED matrix saying, 'Are you there?'."

The Astro Pi units also have a special metal case that makes them — after a few other tweaks to the Raspberry Pi itself — spaceworthy, and we'll talk much more about that in Chapter 2, *High-flyers*. A whole host of experiments designed by British schoolchildren went up with the Astro Pi computers for Tim Peake to try, and the data from those experiments, which all made use of the Sense HAT, was sent back down to Earth.

The Sense HAT is ideal for many applications thanks to its inventive use of sensors and the code that controls it, and in this updated edition of *Experiment with the Sense HAT*, we hope to inspire you to create some cool projects of your own.

The Sense HAT costs £29/$30 and can be purchased from any Raspberry Pi Approved Reseller listed on the Raspberry Pi website (**rpimag.co/sensehat**).

Figure 1-3 The Sense HAT is quite small, but packs a large number of sensors and other features

Chapter 2
High-flyers

British ESA astronaut Tim Peake wasn't the only Brit aboard the International Space Station in 2015. Let's look at how Raspberry Pi first got into the sky and what it means for future generations.

On 15 December 2015, astronaut Tim Peake made history as the first British ESA astronaut to visit the International Space Station (ISS). For the next six months, he achieved most children's dreams as he lived and worked 400 kilometres above the Earth, carrying out a comprehensive science programme during a mission called Principia.

Tim's role was to run experiments in the unique environment of space and try new technologies that may become crucial when humans begin to visit other planets, such as Mars. While aboard the ISS, Tim also conducted a spacewalk to repair the station's power supply and helped dock two spacecraft.

But he was not alone: aside from living with five international colleagues, all of whom had spent years training for their difficult roles, Tim was greeted by another Brit, one set to accompany him for the duration of his time away from Earth. That extra 'colleague' was, of course, the British-made Raspberry Pi. On 6 December 2015, two Raspberry Pis were flown skywards on a Cygnus cargo freighter, departing a little over a week ahead of Tim. By the time Tim set off, the computers were already waiting on board the ISS.

Raspberry Pi's space adventure is referred to as the Astro Pi mission, and the computers as Astro Pi computers. The Raspberry Pi inside each Astro Pi unit was equipped with an add-on sensor board called the Sense HAT and placed inside a cutting-edge aerospace case built to withstand any conditions space would throw at it. As well as allowing the Raspberry Pis to measure the ISS's environment, follow its journey through space, and pick up the Earth's magnetic field, the mission gave schoolchildren the opportunity to have their code run in space for the first time. And that, said Tim, proved to be the most exciting thing of all.

"[Astro Pi] has got a great sensor suite with temperature, pressure, humidity sensors, all sorts of things on it," he told BBC Television's *The One Show* following a final pre-departure press conference on 6 November 2015. "So, the schoolkids basically coded programs that I'm going to run on board the Space Station, and this Astro Pi is going to be in various different modules running an experiment each week. I'm going to send down the data so that, during the mission, they can see [it] — see what they've managed to achieve — and if they need to modify the code, they can send it back up to me."

Figure 2-1 Tim Peake (far right) attends the final press conference before launch

8 · Chapter 2 · High-flyers

Astro Pi was the brainchild of the UK Space Agency and the Raspberry Pi Foundation — although, according to David Honess, the Foundation's Education Resource Engineer, it was also "a case of being in the right place at the right time." Libby Jackson, the UK Space Agency's Astronaut Flight Education Programme Manager, was looking at ways to encourage children to think about technological applications for space and the ISS. "When I was applying for my current role, the candidates were asked to prepare an idea for an activity that could inspire kids, and at the time, I knew about Raspberry Pi," she said. "I didn't take that idea to the interview because I didn't know enough, and I was afraid I'd be asked questions I couldn't answer."

The idea remained with Libby, and when she was talking with UKspace — the UK space industry's trade association — she confessed that she couldn't shake the idea of having fun with Raspberry Pi. "As it happened, someone mentioned that they had been talking to Eben Upton, the CEO of Raspberry Pi, and so had a point of contact. A meeting was quickly set up," Libby explained.

Momentum began building steadily. At the time, David had just begun working with the Raspberry Pi Foundation, and Eben had sent a casual email asking if anyone fancied accompanying him to a meeting with Airbus Defence and Space. David volunteered and found that Dr Stuart Eves, Airbus' Lead Mission Concepts Engineer, was a passionate advocate of the Raspberry Pi. This resulted in the Raspberry Pi Foundation meeting up with Libby at the UK Space Agency, where "Tim Peake's mission was [put] on the table...".

A decision to harness the prospects of that mission as much as possible was soon made, and the idea of setting up a challenge to engage schools was seized upon. The belief was that it would encourage schoolchildren to become more interested in space, opening their eyes to its employment possibilities. "The bottom line is that the UK space industry wants to ensure that there are enough people in the future to hire to carry on doing what they are doing," David explained. "And we feel this is part of the answer."

Once the go-ahead had been given, it was time to work out how the project would run. For Libby, the aim was to attach as much as possible to the Raspberry Pi: "I knew the history of getting education payloads on the ISS," she explained, hinting at the difficulties. The tight schedule they had to work with also posed a problem; the Astro Pi mission was being put together around a year before the flight was expected to take off, so there was never going to be enough time to invite children to come up with an experiment and make it fly. "We turned things on their head and said, 'If we fly the hardware as it exists and ask the kids what we should do with it, that will help in terms of time'," Libby continued. "It seemed the perfect solution."

Astro Pi up close

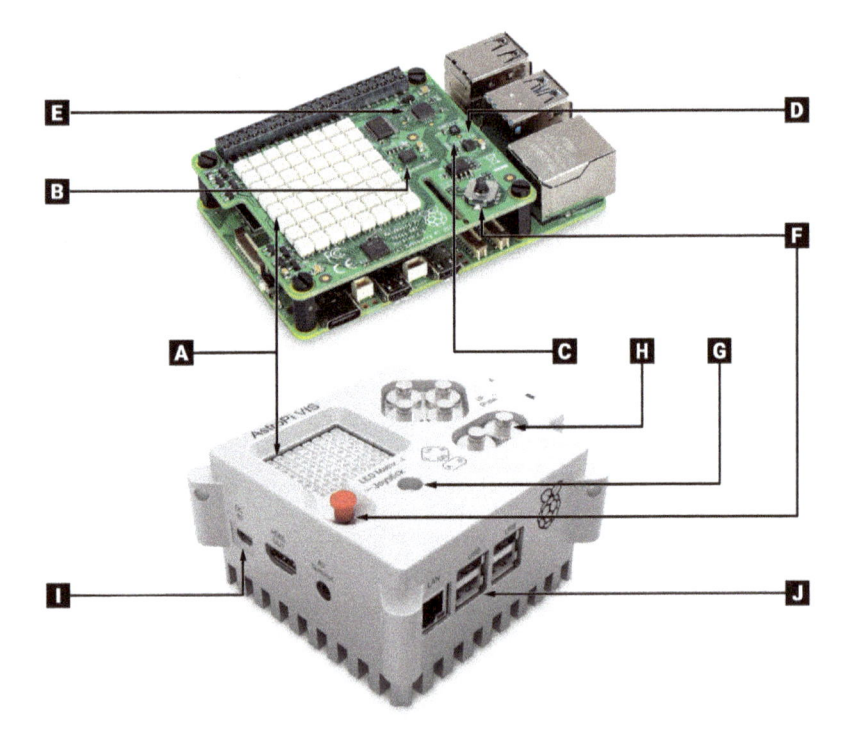

The most noticeable part of the Astro Pi assembly is the 8×8 RGB LED matrix on the Sense HAT (**A**). It has a 60fps refresh rate and 15-bit colour resolution.

The all-in-one gyroscope, accelerometer, and magnetometer (**B**) measures the orientation of objects, the increases in speed, and the strength and direction of a magnetic field. A temperature and humidity sensor (**C**) not only measures hot and cold but also the amount of water vapour in the air; it can be used to detect whether a person is standing close by, for instance. The barometric pressure sensor (**D**) measures the force exerted by small molecules in the air.

The graphics are driven by a microcontroller (**E**).

To enable the astronauts to navigate the screen, there is a five-button joystick (**F**) that allows for up, down, left, and right movements, as well as selection when pushing down, like a button, until it clicks.

A hole in the case (**G**) allows air to enter the Astro Pi device, where it can be detected by the sensors and tested. The actual casing (**H**) is made from aerospace-grade aluminium and is said to have cost £3000 to make. As well as the various special components on the Sense HAT, the usual functions of the Raspberry Pi are used, including the power socket (**I**). The other connections on the Raspberry Pi, from USB sockets to LAN, are also available (**J**).

ON THE CASE

The casing for the Astro Pi is possibly Raspberry Pi's biggest achievement (aside from making the Raspberry Pi computer itself, that is). Large and chunky, it had to adhere to the regulations stipulated by the European Space Agency, which means it must be as safe as a child's toy.

All the edges were inspected to ensure they were not sharp, with testers running their gloved hands over the casing many times to check for potential drag. The heat generated by a Raspberry Pi computer also needed to be conducted away via thermal radiation, so the casing has many pins, each of which can remove 0.1 watts of heat.

The Raspberry Pi is in contact with the case to aid in the heat removal, with tests showing that it will not get hotter than 32−13°C below the cut-off temperature. "There was no aesthetic consideration in designing the case," said David Honess, the Raspberry Pi Foundation's Education Resource Engineer, "but it does look awesome."

Mark ii hardware

In December 2021, after two years of secret development, upgraded Astro Pi VIS and Astro Pi IR computers arrived on the ISS. They were first used to run Mission Zero and Mission Space Lab programs as part of the European Astro Pi Challenge 2021/22.

They were named after the two inspirational European scientists Nikola Tesla and Marie Skłodowska-Curie by Mission Zero participants in 2022.

Pi in the sky

This was not the first time a bare-bones computer had gone into space (and it was not, incidentally, Raspberry Pi's debut either, given Dave Akerman's efforts to strap Raspberry Pi computers to high-altitude balloons and take snapshots from the edge of space). But while Arduinos were the first to boldly go where no other widely accessible device had gone before — onto satellites orbiting the Earth — Astro Pi was created to be different. "Never before have we had a situation where the crew on the Space Station are using the same machine as your kids," David said.

The response from schoolchildren amazed everyone, not only in the quantity of entries but in their quality. There were stories of children coding during their lunch breaks and working on their projects after school. The chance to have their code in space was proving to be a great motivator, and narrowing the experiments down to just seven winners proved tricky. "It came down to the completeness of the ideas and the quality of the coding," Libby revealed. "The things the kids came up with are far more creative than adults."

The winners were certainly impressive. The Cranmere Code Club at Cranmere Primary School, run by teacher (and Raspberry Pi magazine writer) Richard Hayler, tested the humidity surrounding the Astro Pi to determine the whereabouts of the ISS crew. Since a fluctuation in humidity in space is a possible indicator that an astronaut is nearby, the team was able to program the Astro Pi to detect changes in humidity on board and deliver a message to the crew on its 8×8 LED screen, taking a photo with the camera in the hopes of capturing them in action. "They are looking to

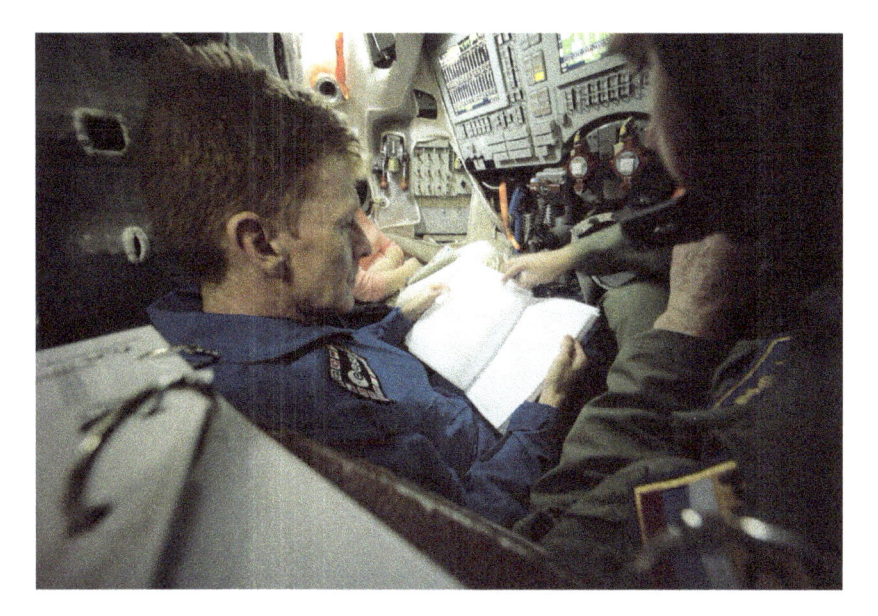

Figure 2-2 Tim during training in the Soyuz TMA simulator

see if humidity is a good indicator of the presence of the crew near the Astro Pi," David explained.

SpaceCRAFT was equally ingenious, with Hannah Belshaw from the Cumnor House Girls School suggesting plugging the output from the Astro Pi sensors into Minecraft so that the environmental measurements could be represented in the game. "SpaceCRAFT logs all sensors to fill a massive CSV file, and it works with code on the ground that plays it back in Minecraft," David said. Hannah dressed in a spacesuit to appear alongside Tim during his BBC interview.

A particular favourite among those involved with Astro Pi was Flags, created by Thirsk School under the watch of teacher Dan Aldred. The program uses telemetry data and the Astro Pi's real-time clock to determine the ISS's location. It searches its database to find the relevant flag and displays an image of it on the LED matrix with a phrase in the local language.

"It's lovely because the children have looked and thought about where the astronauts are in the world," Libby said. David agreed. "The crew will like it," he added. "The kids learned a lot about geography, and they made the code recognise the boundaries of different countries. If it's above the sea, it shows a twinkly blue or green pattern."

Watchdog, by Kieran Wand at Cottenham Village College, made good use of the Astro Pi sensors by measuring the temperature, pressure, and humidity on board the ISS, raising the alarm if they moved outside acceptable parameters. Trees by EnviroPi — a team at Westminster School — pointed the NoIR camera on the Astro Pi out of a window to take images of the ground, from which it could produce a Normalised Differentiated Vegetation Index (a measure of plant health).

Radiation, by the team Arthur, Alexander, and Kiran, overseen by Dr Jesse Peterson at Magdalen College School, used the Camera Module on the Astro Pi to detect radiation by measuring the intensity of specks of light. Believing there is always time for fun, Lincoln UTC's Team Terminal produced a suite of reaction games with teacher Mark Hall, as well as a menu that the astronauts could use to select whichever one they fancied playing at the time.

Get involved

The European Astro Pi Challenge offers young people the amazing opportunity to conduct scientific investigations in space by writing computer programs that run on Raspberry Pi computers aboard the International Space Station.

Find out more and sign up at the Astro Pi website (**astro-pi.org**).

MISSION SPACE LAB
Mission Space Lab offers teams of young people the chance to run scientific experiments on board the ISS.

Find guidelines and a step-by-step guide at **rpimag.co/missionspacelab**.

Figure 2-3   The International Space Station

MISSION ZERO

Mission Zero offers young people the opportunity to run their code in space! Write a simple program to take a reading from the colour and luminosity sensor on an Astro Pi aboard the ISS and use it to set the background colour in a personalised image for the astronauts to see as they go about their daily tasks.

Read the guidelines and the step-by-step guide at **rpimag.co/missionzero**.

> **SEE THE ISS WITH YOUR RASPBERRY PI**
>
> One of the great things about the ISS is that you can see it with your own eyes, without the aid of a telescope. The trick is knowing where it is in the sky at any given time, and there are apps and websites which allow you to follow it as it orbits 400 kilometres above Earth, such as **spotthestation.nasa.gov**.

Tim's role

Tim was able to move between these experiments via an app on the Astro Pi called the Master Control Program (a nod to the 1982 movie *Tron*). Fortunately, he didn't have to keep checking it; the programs could run automatically. "There is a clock icon which will run program X for a set period," David explained. "It ensures the programs run for the right amount of time."

Schedules defining how many seconds each experiment should run for were specified. "He can use the joystick to go down the different programs, and if he wants to run one, then he can press the 'right' button, which shows an arrow on the screen and then starts that program," David said. "The results are written to the SD card, and they go into a folder called **Transfer**, which Tim can copy and send down to us."

Figure 2-4   Tim tries on a spacesuit

Tim's work included experiments of his own, away from the Astro Pi. One involved studying metals using the on-board electromagnetic levitator, a furnace which heats metals to 2100°C and rapidly cools them in a gravity-free environment. The removal of gravity allows for more accurate observation of the fundamental properties of different metals and alloys, as well as their rates of cooling. He also looked at organisms placed on the exterior of the ISS to see how lack of oxygen, extreme temperature changes, and radiation affected them.

Perhaps most importantly, Tim's research included measuring brain pressure in space. There has long been concern that space exploration (and time on the ISS) can affect astronauts' vision, as low gravity allows blood to rise, increasing brain pressure and pushing on the back of the eyes. Tim's work is helping researchers at the University Hospital Southampton NHS Foundation Trust better understand the open fluid links between the brain and the ear, which could be a better way to test astronaut health.

But where did that leave the educational part of Tim's mission? "In the official world, Tim will have four hours of education activity time per expedition," Libby explained. With Tim on Expedition 46 and 47, that equates to eight hours total. That doesn't sound a lot, and Libby admitted it isn't — "in space everything floats, so we usually work out how long it will take to do something on Earth and triple the time" — but Tim was brilliantly committed to ensuring that the mission was fun for the next generation of children. To that end, he wanted them to get the most out of it. "He will spend a lot of Saturday afternoons working on education projects," Libby said. "Astro Pi is one of our flagship education programmes, and we're looking forward to it. Education is going to be very important in Tim's mission."

The Astro Pi missions, which are still taking place today, could well represent a turning point for the UK space industry. "Only a small number of people can be an astronaut, and that is what kids think about," David said. "They also see space as abstract and only associate it with NASA. But we are showing the various roles and possibilities [that exist within the field]. We're calling it the 'Tim Peake effect' [akin to the 'Apollo effect' in the USA in the 1960s and 1970s, which boosted interest in science and engineering], and we hope that [one day] the UK will have a booming space industry. It's a bold aim, but it's everybody's hope."

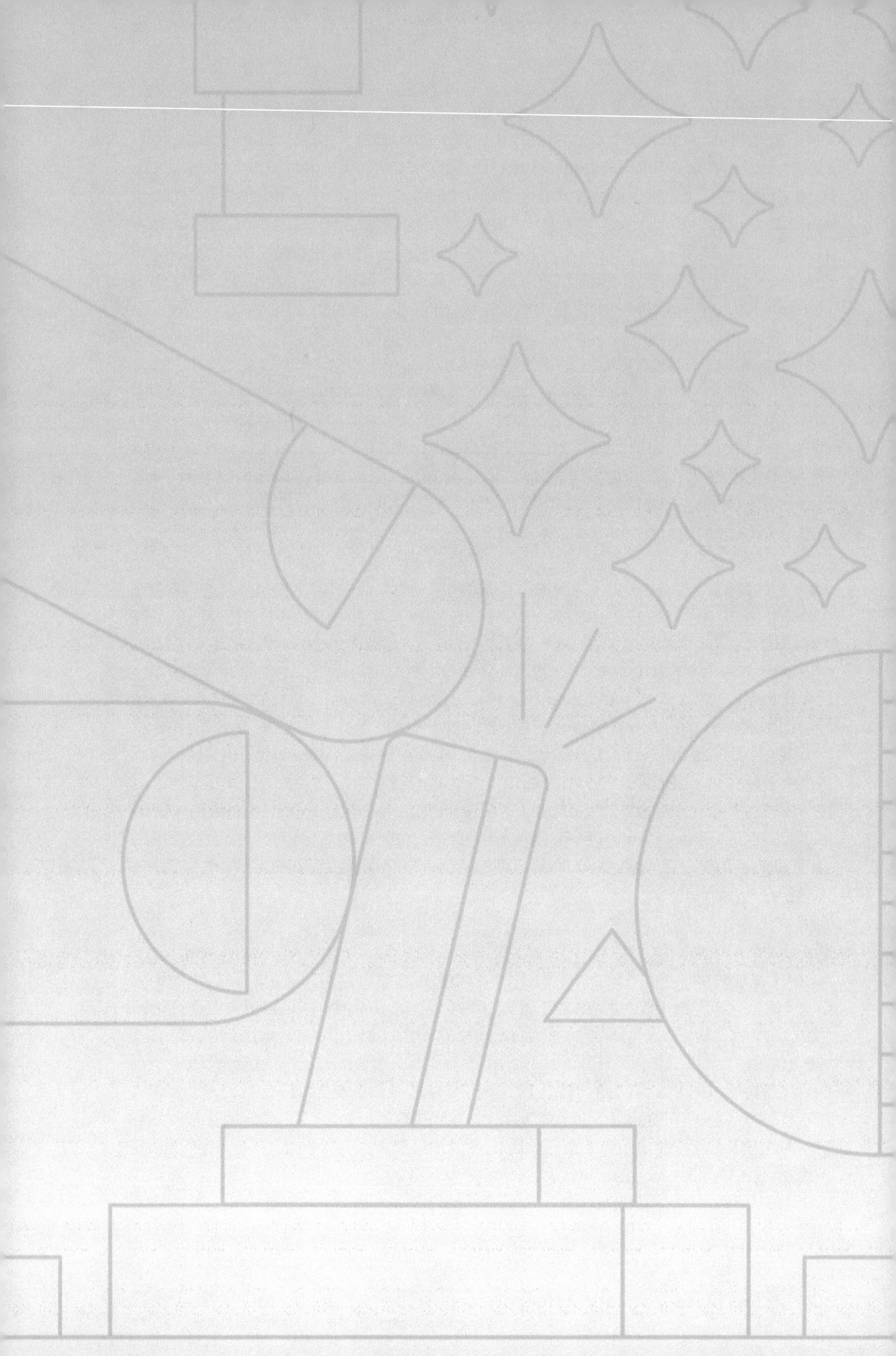

Chapter 3

Get started with the Sense HAT

Now that you know all about the Sense HAT and the part it plays in Astro Pi, it's time to learn how you can actually use one for your own stellar projects

You don't need to be in space to use the Sense HAT: it works down on Earth as well! Once you've managed to get your hands on one, you'll probably want to start using it, which is where this chapter comes in handy.

Assemble the Sense HAT

If you have not yet installed the Sense HAT, now is the time to do so. First, make sure your Raspberry Pi is shut down and unplugged from the mains, and that all other cables have been disconnected. The Sense HAT comes in an anti-static bag, along with the following fixtures and fittings. Make sure that these are all present before proceeding:

- 1 × GPIO pin extension header
- 4 × hexagon standoffs
- 8 × M2.5 screws

Figure 3-1 shows how it all fits together.

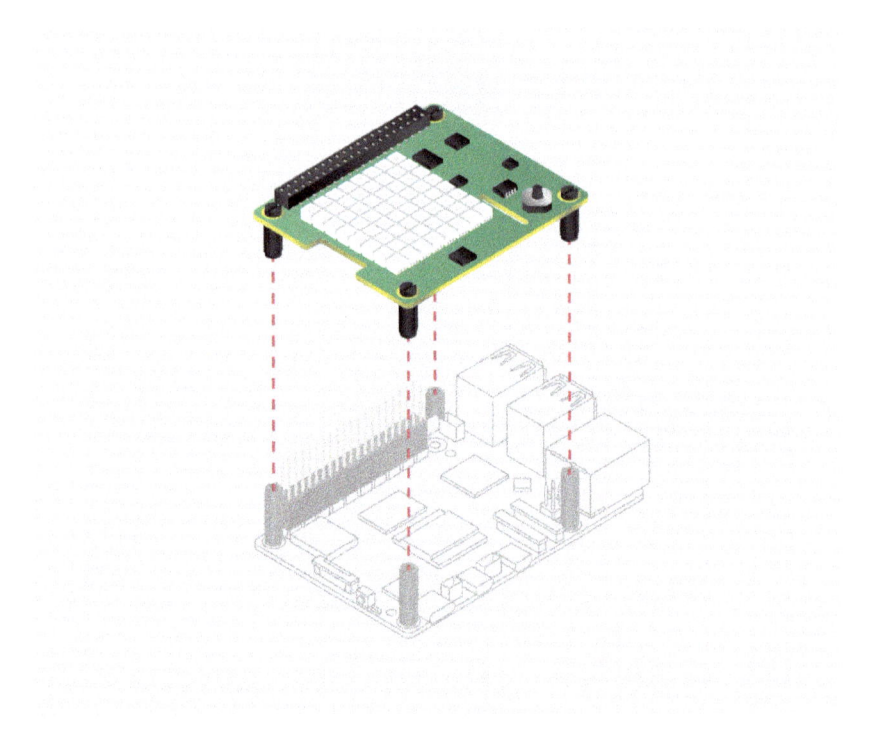

Figure 3-1 Attaching the Sense HAT to a Raspberry Pi

1. Put the GPIO extension header block onto the Raspberry Pi's GPIO pins
2. Screw the hexagon standoffs to the Raspberry Pi by threading the screws through from the bottom and turning the standoffs between your finger and thumb
3. Insert the Sense HAT into the GPIO pin extension header; the corner holes should align with the hexagon standoffs
4. Put the remaining screws through from the top

Use a small Phillips screwdriver to tighten each corner standoff individually. They don't need to be especially tight — just enough to ensure that the HAT doesn't become loose.

Now boot your Raspberry Pi to the Raspberry Pi OS desktop and start Thonny: click the Raspberry Pi menu and select **Programming > Thonny**.

Introducing the Thonny Python IDE

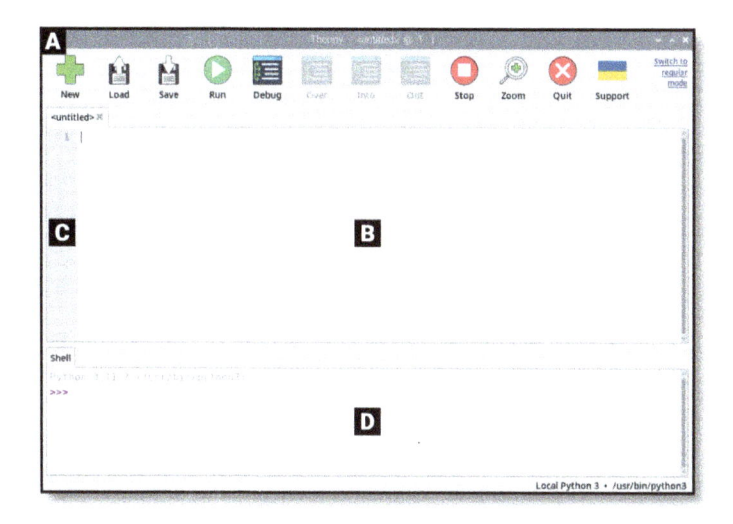

A Toolbar

B Script area

C Line numbers

D Python shell

Thonny's 'Simple Mode' interface uses a bar of friendly icons (**A**) as its menu, allowing you to create, save, load, and run your Python programs, as well as test them in various ways. The script area (**B**) is where your Python programs are written, and is split into a main area for your program and a small side margin for showing line numbers (**C**). The Python shell (**D**) allows you to type individual instructions which are then run as soon as you press the **ENTER** key, and also provides information about running programs.

> **THONNY MODES**
>
> Thonny has two main versions of its interface: 'Regular Mode' and a 'Simple Mode', which is better for beginners. This chapter uses Simple Mode, which is loaded by default when you open Thonny from the **Programming** section of the Raspberry Pi menu.

Hello, Sense HAT

For your first trick, you'll display text on the HAT's LED matrix. This program contains two crucial lines of code that import the Sense HAT library and create a **sense** object to represent the HAT:

```
from sense_hat import SenseHat
sense = SenseHat()
```

You can find full documentation for the Sense HAT Python module at **rpimag.co/sensehatmodule**. The next line makes the Sense HAT display some text:

```
sense.show_message("Hello, Sense HAT!")
```

You can easily change the message inside the quotation marks to display your own text, but you can also do much more than that: for example, we can expand the **sense.show_message** command to include some extra parameters that will change the behaviour of the message.

The following program will display the text "**Astro Pi is awesome!**" more slowly, with the text in yellow **[255, 255, 0]** and the background in blue **[0, 0, 255]**:

```
from sense_hat import SenseHat
sense = SenseHat()
sense.show_message("Astro Pi is awesome!", scroll_speed=0.05,
                    text_colour=[255, 255, 0],
                    back_colour=[0, 0, 255])
```

You could also make the message repeat by using a **while** loop:

```
from sense_hat import SenseHat
sense = SenseHat()
while True:
    sense.show_message("Astro Pi is awesome!!",
                        scroll_speed=0.05,
                        text_colour=[255,255,0],
                        back_colour=[0,0,255])
```

Now we've made our first program, we should save it. Click **File > Save As**, give your program a name like **loop_text.py**, then press **F5** to run it. Easy!

The LED matrix can also display a single character, rather than an entire message, using the `sense.show_letter` function, which has the same optional parameters:

Parameter	Effect
`scroll_speed`	This parameter affects how quickly the text moves on the screen. The default value is 0.1. The bigger the number, the slower the speed.
`text_colour`	This alters the text colour and is specified as three values: red, green, and blue. Each value can be between 0 and 255, so `[255,0,255]` would be red + blue = purple.
`back_colour`	This parameter alters the colour of the background and is also specified as values for red, green, and blue.

The Sense HAT emulator

If you don't have a Sense HAT, you can use the Sense HAT emulator, which can emulate every part of the Sense HAT except the colour sensor. To install the Sense HAT emulator, click the Raspberry Pi icon, choose **Accessories > Terminal**, and run the following code:

`sudo apt -y install sense-emu-tools`

Python code written for a real Sense HAT runs on the emulator, and vice-versa, with only one change: if you're using the Sense HAT emulator, you'll need to change the line `from sense_hat import SenseHat` to `from sense_emu import SenseHat` instead. If you want to then run them on a physical Sense HAT again, just change the line back!

When you run a Python program with **sense_emu** instead of **sense_hat**, it will start the emulator if it's not running already (see **Figure 3-2**). The emulator displays the Sense HAT LEDs and includes sliders that let you set the sensor values; it also includes buttons that let you emulate the joystick.

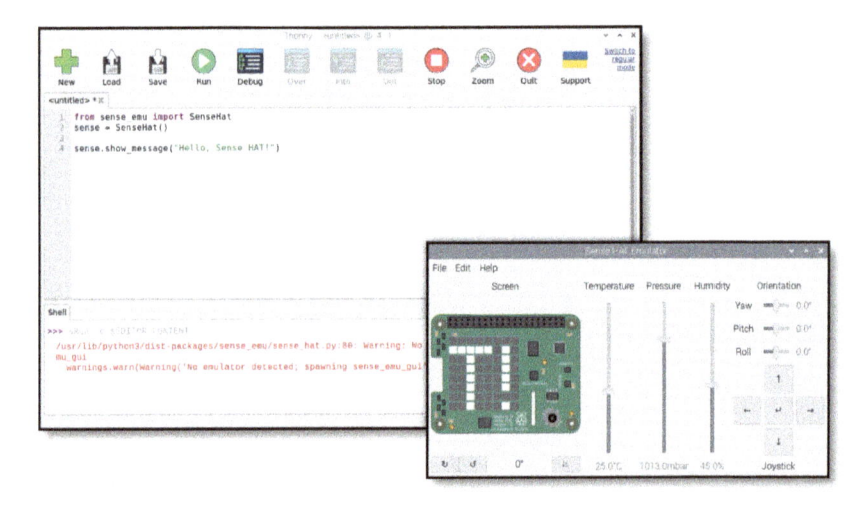

Figure 3-2 The Sense HAT emulator

Displaying images

Of course, the LED matrix can display more than just text. We can control each LED individually to create our own images, and there are a couple of different ways we can accomplish this. The first approach is to set pixels (LEDs) individually; we can do this using the **sense.set_pixel()** command. First, we need to be clear about how we describe each pixel.

The Sense HAT uses a coordinate system, and the numbering begins at 0, not 1. The origin is in the top left rather than the bottom left, as you may be used to. In Thonny, click **New** and try the following program:

```
from sense_hat import SenseHat
sense = SenseHat()
sense.set_pixel(0, 1, [0, 0, 255])
sense.set_pixel(7, 4, [255, 0, 0])
```

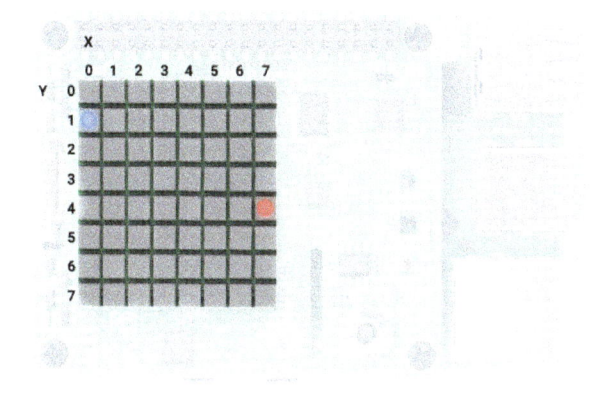

Figure 3-3   The Sense HAT's LED matrix uses a handy coordinate system

As you can see in **Figure 3-3**, setting pixels individually works, but it gets rather complex when you want to set lots of them. There is another option, though: `sense.set_pixels`.

This is quite straightforward — we just give a list of colour values for each pixel. We could enter:

sense.set_pixels([[255, 0, 0], [255, 0, 0],
                  [255, 0, 0], [255, 0, 0],...])

...but this would take ages. Instead, you can use some variables to define your colour palette. In this example, we're using the colours of the rainbow:

```
r = [255, 0, 0]
o = [255, 127, 0]
y = [255, 255, 0]
g = [0, 255, 0]
b = [0, 0, 255]
i = [75, 0, 130]
v = [159, 0, 255]
e = [0, 0, 0]   # e is for empty
```

Experiment with the Sense HAT, 2nd Edition · 25

We can then describe our matrix by creating a 2D list of colour names:

```
image = [ e,e,e,e,e,e,e,e,
          e,e,e,r,r,e,e,e,
          e,r,r,o,o,r,r,e,
          r,o,o,y,y,o,o,r,
          o,y,y,g,g,y,y,o,
          y,g,g,b,b,g,g,y,
          b,b,b,i,i,b,b,b,
          b,i,i,v,v,i,i,b ]
```

Once you have the colour and image variables, you can then simply call them by adding:

`sense.set_pixels(image)`

...but don't forget to start your listing with:

```
from sense_hat import SenseHat
sense = SenseHat()
```

Click **File > Save As**, give your program a name (e.g. **rainbow.py**), then press **F5** to run. What will you display on your Sense HAT?

Setting orientation

So far, all our text and images have appeared the same way up, assuming that the HDMI port is at the bottom. However, this may not always be the case (especially in space), so you may want to change the orientation of the matrix. To do this, you can use the `sense.set_rotation()` method. Inside the brackets, enter one of four angles (0, 90, 180, or 270).

To rotate your screen by 180 degrees, you'd use this line:

`sense.set_rotation(180)`

When used in the rainbow program, it would look like this:

```
from sense_hat import SenseHat
sense = SenseHat()

r = [255, 0, 0]
o = [255, 127, 0]
y = [255, 255, 0]
g = [0, 255, 0]
b = [0, 0, 255]
i = [75, 0, 130]
v = [159, 0, 255]
e = [0, 0, 0]

image = [ e,e,e,e,e,e,e,e,
         e,e,e,r,r,e,e,e,
         e,r,r,o,o,r,r,e,
         r,o,o,y,y,o,o,r,
         o,y,y,g,g,y,y,o,
         y,g,g,b,b,g,g,y,
         b,b,b,i,i,b,b,b,
         b,i,i,v,v,i,i,b ]

sense.set_pixels(image)
sense.set_rotation(180)
```

Click **File > Save As**, give your program a name (e.g. **rainbow_flip.py**), then press **F5** to run.

You could also create spinning text using a **for** loop. In Thonny, click **New** and type in the following code:

```
from sense_hat import SenseHat
import time
sense = SenseHat()

sense.show_letter("J")

angles = [0, 90, 180, 270, 0, 90, 180, 270]
for r in angles:
    sense.set_rotation(r)
    time.sleep(0.5)
```

This program displays the letter 'J' and then sets the rotation to each value in the angles list with a 0.5-second pause. Click **File > Save As**, give your program a name (e.g. **spinning_j.py**), then press **F5** to run.

You can also flip the image on the screen, either horizontally or vertically, using `sense.flip_h()` or `sense.flip_v()`. With this example, you can create a simple animation by flipping the image repeatedly:

```
from sense_hat import SenseHat
import time

sense = SenseHat()

w = [150, 150, 150]
b = [0, 0, 255]
e = [0, 0, 0]

image = [ e,e,e,e,e,e,e,e,
         e,e,e,e,e,e,e,e,
         w,w,w,e,e,w,w,w,
         w,w,b,e,e,w,w,b,
         w,w,w,e,e,w,w,w,
         e,e,e,e,e,e,e,e,
         e,e,e,e,e,e,e,e,
         e,e,e,e,e,e,e,e ]

sense.set_pixels(image)

while True:
    time.sleep(1)
    sense.flip_h()
```

Click **File > Save As**, give your program a name (e.g. **eyes.py**), then press F5 to run.

Sensing the environment

The Sense HAT has several environmental sensors which enable it to detect various conditions, including pressure, temperature, and humidity. We can collect these readings using three simple methods:

sense.get_temperature()
    This will return the temperature in Celsius

sense.get_pressure()
    This will return the pressure in millibars

sense.get_humidity()
    This will return the humidity as a percentage

Using these, you could create a simple scrolling text display to keep people informed about current conditions. Create a new program and try this code:

```python
from sense_hat import SenseHat
sense = SenseHat()

while True:
    t = sense.get_temperature()
    p = sense.get_pressure()
    h = sense.get_humidity()
    
    t = round(t, 1)
    p = round(p, 1)
    h = round(h, 1)
    
    msg = f"Temperature={t}, Pressure={p}, Humidity={h}"
    
    sense.show_message(msg, scroll_speed=0.05)
```

Click **File > Save As**, give your program a name (e.g. **env.py**), then press F5 to run.

You could also use colour to let astronauts know whether conditions are within sensible ranges. According to some online documentation, the International Space Station maintains the following environmental conditions:

- Temperature: 18.3–26.7 degrees Celsius
- Pressure: 979–1027 millibars
- Humidity: around 60%

You could use an `if` statement in your code to check these conditions and set a background colour for the scroll:

```
if t > 18.3 and t < 26.7:
    bg = [0, 100, 0] # green
else:
    bg = [100, 0, 0] # red
```

Your complete program would look like this:

```
from sense_hat import SenseHat
sense = SenseHat()

while True:
    t = sense.get_temperature()
    p = sense.get_pressure()
    h = sense.get_humidity()

    t = round(t, 1)
    p = round(p, 1)
    h = round(h, 1)

    if t > 18.3 and t < 26.7:
        bg = [0, 100, 0]   # green
    else:
        bg = [100, 0, 0]   # red

    msg = f"Temperature={t}, Pressure={p}, Humidity={h}"

    sense.show_message(msg, scroll_speed=0.05, back_colour=bg)
```

Click **File > Save As,** give your program a name (e.g. **scrolling_env.py**), then press **F5** to run.

Detecting movement

The Sense HAT also has a set of sensors that can detect movement. It has an IMU (inertial measurement unit) chip containing a gyroscope for detecting which way up the board is, an accelerometer for detecting movement, and a magnetometer for detecting magnetic fields.

Before experimenting with motion-sensing, you should understand three key terms used when talking about the three axes of motion: *pitch*, *roll*, and *yaw* (see **Figure 3-4**). It's helpful to visualise these concepts using the example of an aeroplane: pitch is like a plane taking off or diving, roll is like a plane spinning, and yaw is like a plane's nose steering left and right.

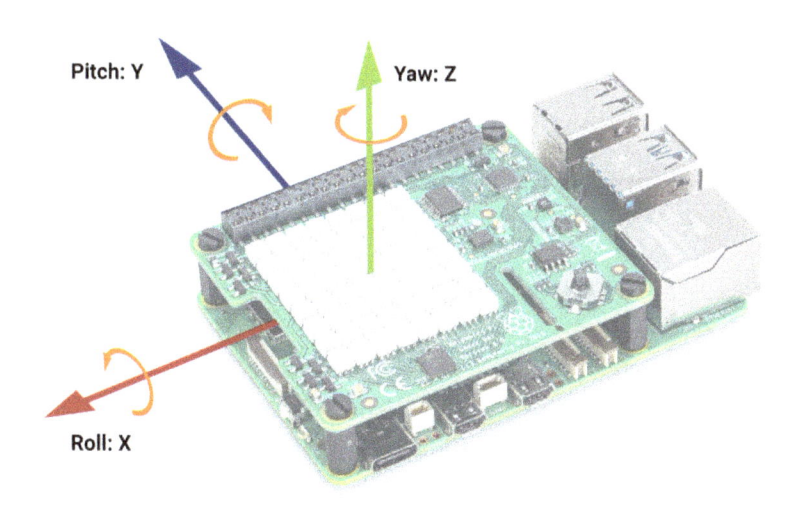

Figure 3-4   Roll, pitch, and yaw

You can find out the orientation of the Sense HAT using the **sense.get_orientation()** method:

pitch, roll, yaw = sense.get_orientation().values()

This will produce the three orientation values (measured in degrees) and store them as the three variables **pitch**, **roll**, and **yaw**. The **.values()** obtains the three values so that they can be stored separately.

You can explore these values with a simple program:

```python
from sense_hat import SenseHat

sense = SenseHat()

while True:
    pitch, roll, yaw = sense.get_orientation().values()
    print(f"pitch={pitch}, roll={roll}, yaw={yaw}")
```

Click **File > Save As**, give your program a name (e.g. **orientation.py**), then press **F5** to run.

When using the movement sensors, it is important to poll them often in a tight loop. If you poll them too slowly, for example with **time.sleep(0.5)** in your loop, you will see strange results. This is because the code behind needs lots of measurements to successfully combine the data coming from the gyroscope, accelerometer, and magnetometer. For this reason, we're using **print()** to display the output in Thonny itself instead of **sense.show_message()**, which would slow down the polling.

Another way to detect orientation is to use the **sense.get_accelerometer_raw()** method, which tells you the amount of g-force acting on each axis. If any axis has ±1g, then you know that axis is pointing downwards.

In this example, the amount of gravitational acceleration for each axis (x, y, and z) is extracted and then rounded to the nearest whole number:

```python
from sense_hat import SenseHat

sense = SenseHat()

while True:
    x, y, z = sense.get_accelerometer_raw().values()

    x=round(x, 0)
```

```
y=round(y, 0)
z=round(z, 0)

print(f"x={x}, y={y}, z={z}")
```

Click **File > Save As**, give your program a name (e.g. **acceleration.py**), then press **F5** to run. As you turn the screen, you should see the values for x and y change between -1 and 1. If you place the Raspberry Pi flat or turn it upside down, the z-axis will be 1 and then -1.

If we know which way round the Raspberry Pi is, we can use that information to set the orientation of the LED matrix. First, you would display something on the matrix, then continually check which way round the board is and use that to update the orientation of the display:

```
from sense_hat import SenseHat

sense = SenseHat()

sense.show_letter("J")

while True:
    x, y, z = sense.get_accelerometer_raw().values()

    x = round(x, 0)
    y = round(y, 0)

    if x == -1:
        sense.set_rotation(90)
    elif y == 1:
        sense.set_rotation(0)
    elif y == -1:
        sense.set_rotation(180)
    else:
        sense.set_rotation(270)
```

Click **File > Save As**, give your program a name (e.g. **rotating_letter.py**), then press **F5** to run.

In this program you are using an **if**, **elif**, **else** structure to check which way round the Raspberry Pi is. The **if** and **elif** test three of the orientations, and if the orientation doesn't match any of these, then the program assumes it is the "right" way round. By using the **else** statement, we also catch all those other situations, like when the board is at a 45-degree angle or is sitting level.

If the board is only rotated, it will only experience 1g of acceleration in any direction; if we were to shake it, the sensor would experience more than 1g. We could then detect that rapid motion and respond. For this program, we will introduce the **abs()** function, which is not specific to the Sense HAT library and is part of standard Python. **abs()** gives us the size of a value and ignores whether the value is positive or negative. This is helpful because we don't care which direction the sensor is being shaken in, just that it is being shaken:

```python
from sense_hat import SenseHat

sense = SenseHat()

while True:
    x, y, z = sense.get_accelerometer_raw().values()

    x = abs(x)
    y = abs(y)
    z = abs(z)

    if x > 1 or y > 1 or z > 1:
        sense.show_letter("!", text_colour=[255, 0, 0])
    else:
        sense.clear()
```

Click **File > Save As**, give your program a name (e.g. **shake.py**), then press F5 to run. You might find that this is quite sensitive, in which case you could change the value from 1 to a higher number.

We've now covered every function that the Sense HAT is capable of — let's bring it all together in a series of fun projects in the remaining chapters!

Chapter 4

Random sparkles

Create amazing random sparkles with your Sense HAT as you learn how to set random pixels in various colours on its LED display

Before we get into working with the Sense HAT's sensors in later chapters, let's blink some LEDs. First, we'll need to think up some random numbers and use the **set_pixel** function to place a random colour in a random location on the Sense HAT's 8×8 LED display. Open the Thonny editor (click the Raspberry Pi menu and select **Programming > Thonny**); if Thonny is already open, press the **New** button to start a new program.

Next, import the Sense HAT software and create a **sense** object to represent the HAT:

```
from sense_hat import SenseHat
sense = SenseHat()
```

Define your variables

Now you need to define **x** and **y** to choose which pixel on the Sense HAT will light up. Create a variable called **x** and set it equal to a number of your choice, between 0 and 7. This will be the x coordinate of your pixel on the display:

```
x = 4
```

Create another variable called **y** and set it equal to another number between 0 and 7. This will be the y coordinate of your pixel on the display.

```
y = 7
```

To choose the colour of your pixel, think of three numbers between **0** and **255**, then assign them to variables called **r**, **g**, and **b**. These variables will represent the colour of your pixel as amounts of red (**r**), green (**g**), and blue (**b**).

```
r = 19
g = 180
b = 230
```

We first encountered variables in Chapter 3, *Get started with the Sense HAT*, but here's a little more detail on how they work. A variable allows you to store data within a program. Variables have both a name and a value. This variable has the name **animal** and the value **cat**:

```
animal = "cat"
```

This variable has the name **score** and the value **30**:

```
score = 30
```

To create a variable, give it a name and set it equal to a value. The name of the variable always goes on the left, so the following code is wrong:

```
# This code is wrong
30 = score
```

Next, you'll use the **set_pixel** function to place a pixel with your randomly chosen colour at your randomly chosen location on the display.

Set a single pixel on the Sense HAT

The **set_pixel** method takes data in the following order: x coordinate, y coordinate, red, green, blue. Plug the names of your variables into this line of code in the same order:

```
sense.set_pixel(x, y, red_value, green_value, blue_value)
```

Below is what your code should look like. You can choose any x and y values between 0 and 7, and any **r**, **g**, and **b** values between **0** and **255**:

```
from sense_hat import SenseHat
sense = SenseHat()

x = 4
y = 5
r = 19
g = 180
b = 230

sense.set_pixel(x, y, r, g, b)
```

Click **Save**, name your program **sparkles.py**, then press **F5** to run it.

Generate random numbers

So far, you've picked your own random numbers, but you can let the computer choose them by using one of the standard modules in Python. The **random** module creates *pseudo-random* numbers in your code. Three of the functions you'll use frequently are **randint**, **uniform**, and **choice**:

randint
> You can generate random integers between two values using the **randint** function. For example, the following line of code will produce a random integer between 0 and 10 (inclusive).
>
> ```
> from random import randint
> num = randint(0, 10)
> ```

uniform

> If you want a random floating-point number (also called *float*), you can use the **uniform** function. For example, the following line of code will produce a random float that's equal to or greater than 0, but less than 10.
>
> ```
> from random import uniform
> num = uniform(0, 10)
> ```

choice

> If you want to choose a random item from a list, you can use the **choice** function.
>
> ```
> from random import choice
> deck = ['Ace', 'King', 'Queen', 'Jack']
> card = choice(deck)
> ```

Add another **import** line at the top of your program:

```
from random import randint
```

Change your **x** and **y** variables to equal a random number between 0 and 7. Now your program will automatically select a random position on the LED matrix.

```
x = randint(0, 7)
y = randint(0, 7)
```

Run your program again, and you should see another random pixel placed on the Sense HAT's display. It will be the same colour you chose previously.

Next, change your **r**, **g**, and **b** variables to equal a random number between 0 and 255. Now your program will automatically select a random colour.

```
r = randint(0, 255)
g = randint(0, 255)
b = randint(0, 255)
```

Run the program again, and you should see another pixel appear in a random location, this time in a random colour. Run it a few more times, and you should see more of the grid fill up with random pixels!

Add a loop

Rather than running your program over and over by clicking Run (or pressing **F5**), you can add a loop so that it will keep running by itself. You can use the sleep module to pause the program between pixels. To do so, first add another import to the top of your file.

```
from time import sleep
```

Add an infinite loop on the line below the `sense = SenseHat()` statement. Be sure to indent all the lines of code containing your variables and `set_pixel` so that they are within the loop:

```
while True:
    x = randint(0, 7)
    y = randint(0, 7)
    r = randint(0, 255)
    g = randint(0, 255)
    b = randint(0, 255)
    sense.set_pixel(x, y, r, g, b)
```

Python's while loop

You saw `while` loops back in Chapter 3, *Get started with the Sense HAT*, but we didn't go into a lot of detail at the time. The purpose of a while loop is to repeat code over and over while a condition is `True`. This is why while loops are sometimes referred to as condition-controlled loops.

The example below is a while loop that will run forever — an infinite loop. The loop will run forever because the condition is always `True`.

```
while True:
    print("Hello world")
```

The **while** line states the loop condition. The **print** line of code below it is indented slightly to the right. Indentation is created using exactly four spaces and shows that a line of code is inside the loop. Any code inside the loop will be repeated.

An infinite loop is useful in situations where you want to perform the same actions over and over again, for example, checking the value of a sensor. An infinite loop like this will continue to run forever, meaning any lines of code written after the loop will never happen. This is known as *blocking* — the program blocks the execution of any other code.

Add a pause

Add a line of code at the bottom of your program to pause for 0.1 seconds. Make sure that this line is indented, level with the **set_pixel** line, to show that it is inside the loop:

```
sleep(0.1)
```

Try it out

Run the code to see your sparkles in action! Here's the complete program:

```
from sense_hat import SenseHat
from random import randint
from time import sleep

sense = SenseHat()

while True:
    x = randint(0, 7)
    y = randint(0, 7)
    r = randint(0, 255)
    g = randint(0, 255)
    b = randint(0, 255)
    sense.set_pixel(x, y, r, g, b)
    sleep(0.1)
```

Challenge: Better sparkles

Can you make the sparkles change more quickly? Can you make the sparkles appear in pastel colours? (Hint: Normally, you pick colour variable values within the range of 0 to 255; reduce the range and see what happens.)

Try fixing one of the colour values to 0 and see what happens. You can also try making the sparkles disappear, either the same way as they appeared (one pixel per loop) or perhaps by using a wipe-style transition.

Figure 4-1 The LED matrix can display a variety of colours

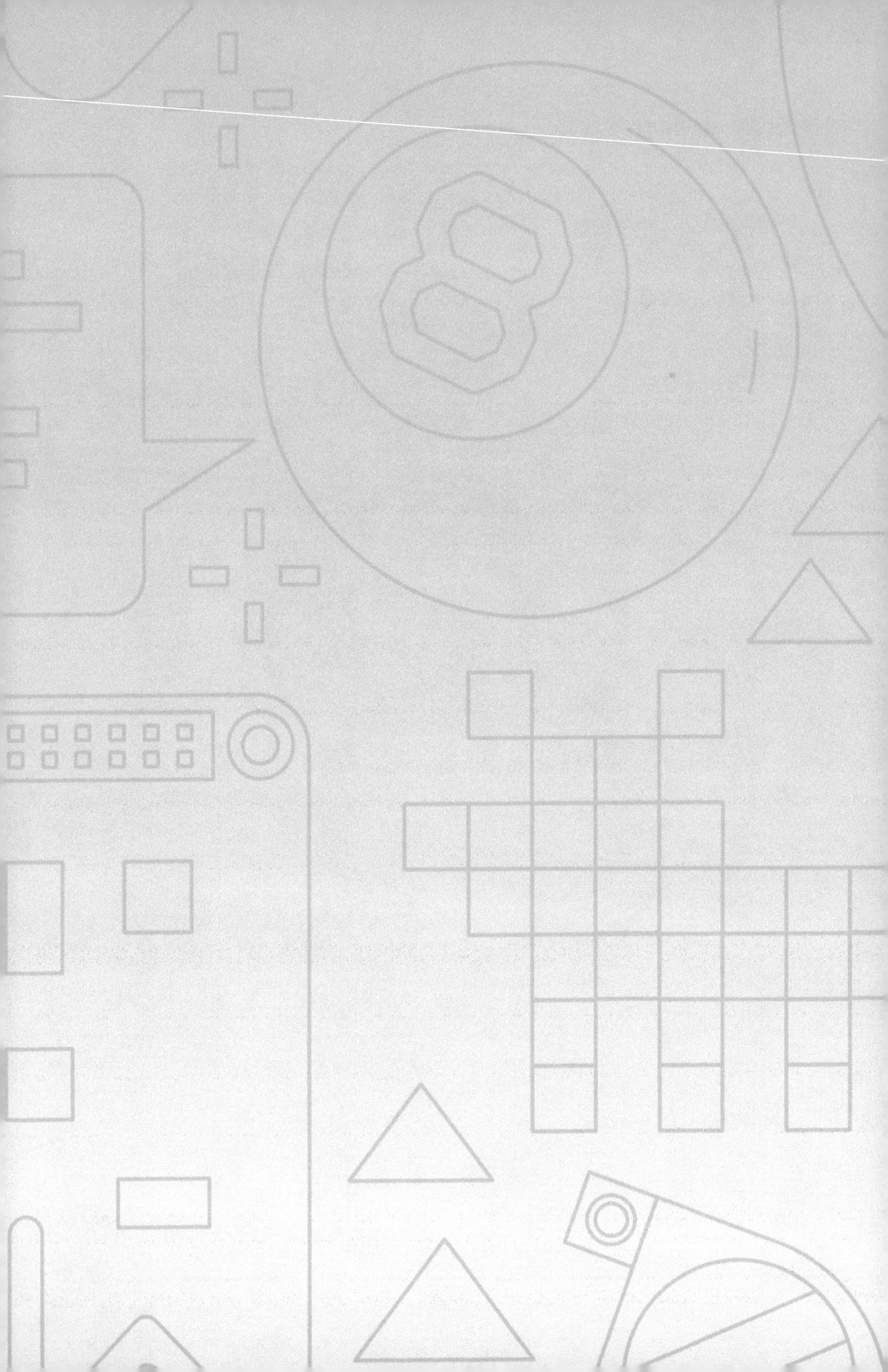

Chapter 5

Make a digital twist on the Magic 8 Ball

Bring the time-honoured tradition of shaken-not-stirred fortunes to the Sense HAT, making use of its built-in motion sensors

In this activity, you will build your own homage to Mattel's Magic 8 Ball using your Raspberry Pi, a Sense HAT, and some Python code. The Magic 8 Ball is a toy in the shape of an oversized eight-ball containing fluid and a 20-sided polyhedron with a response on each face. To use it, you verbally ask a closed question, shake the ball, and read whichever of the 20 answers appears in the window on the bottom as the polyhedron rises to the top of the fluid.

In this tutorial, you will use Thonny to write some code for the Sense HAT. This means you can test your code and fix it as you write it. Open the Thonny editor (click the Raspberry Pi menu and select **Programming > Thonny**); if Thonny is already open, press the **New** button to start a new program.

Once Thonny has loaded, it will open a text editor window in which you can write, save, and test your code. Save the blank file as **magic8ball.py** by clicking **Save**.

Print replies to the screen

A good way to start your Magic 8 Ball program is to create a text version of how it works. Let's think about what you do with a Magic 8 Ball: first, you ask it a question, then you shake the ball, turn it over, and read a reply that it has randomly selected. You will therefore need a list of replies and a way of randomly choosing one and displaying it on the screen.

To start, you need to import the **random** and **time** libraries. Type the following into your **magic8ball.py** text file:

```python
import random
import time
```

Using the **print** function, you can print text to the screen for the person using your program to read. Type:

```python
print("Ask a question")
```

There needs to be a pause before the program responds so that the user can ask a question verbally or mentally. You can use the **time** library to ask the program to sleep for a set amount of time, like this:

```python
time.sleep(3)
```

The program will pause for three seconds. You can change this sleep value to make the pause longer or shorter. Now, create a list of replies that the program could give to the question. Lists can be named in much the same way as variables: for example, **number = [1, 2, 3, 4]**. This list called 'number' has four items in it. Your list will contain strings of text that will be displayed on the screen; these strings will be quite long. To create your list, type:

```python
replies = ['Signs point to yes', 'Without a doubt',
           'You may rely on it']
```

Add as many replies to your list as you like. Make sure to separate each reply with a comma. You can break up your list onto multiple lines, like this, to make it easier to read; however, this is not required for your program to work:

```
replies = [
'Signs point to yes',
'Without a doubt',
'You may rely on it',
'Do not count on it',
'Looking good',
'Cannot predict now',
'It is decidedly so',
'Outlook not so good'
]
```

Finally, an instruction is needed to select an item from the list at random and display it on the screen. You can use the **random** library to do this:

print(random.choice(replies))

Save your code by clicking **Save**, then run your program to test it works by clicking **Run** or pressing **F5**. Your code should resemble that in **Figure 5-1**.

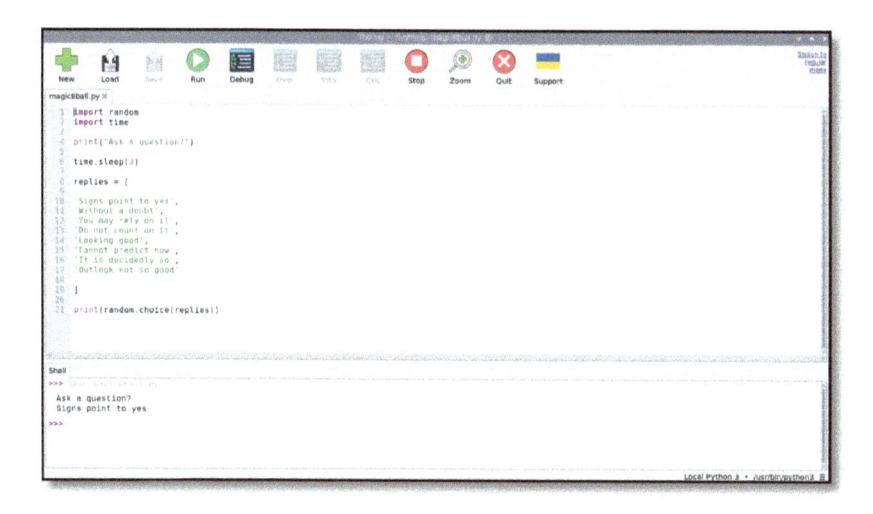

Figure 5-1   magic8ball Python code running in the Thonny shell

Display text on the Sense HAT

Now that you have text outputting to the Python 3 shell window on your screen, you can change the code so that the text scrolls across the LED matrix on your Sense HAT. To do this, you will need to use the **SenseHat** library and replace the print functions with the **show_message** function. Underneath the **import** section of your code, add the following lines:

```
from sense_hat import SenseHat
sh = SenseHat()
```

> **What's in a name?**
>
> In this chapter, we're calling our Sense HAT object **sh** instead of **sense**, in contrast to previous chapters, illustrating that you can use any legal variable name in your own code. In Python, variables can only contain numbers, letters, and an underscore; they must start with a letter or an underscore and cannot start with a number. They are also case-sensitive, meaning **sense** and **senSe** would be two different variables.

Next, replace **print** with **sh.show_message** in your code. There are two places where you will need to do this (see **Figure 5-2**).

To test the code, save your program by clicking **Save**, then run it to check that it works on the Sense HAT by clicking **Run** or pressing **F5**.

You may find that the text is slow to scroll across the LED matrix on your Sense HAT. To speed up the text, you can add **scroll_speed=(0.06)** to your text strings.

Normal Magic 8 Balls work by being physically shaken up. How do you think you would make that happen using the Sense HAT's motion sensors? That's your next programming challenge! If you need a hint, check out the **shake.py** example at the end of "Detecting movement" on page 31. The solution is in the GitHub repository at **rpimag.co/sensebookgit**, but try to work it out yourself before you peek at it!

Figure 5-2 magic8ball Python code adapted to display on the Sense HAT LEDs

Experiment with the Sense HAT, 2nd Edition · 49

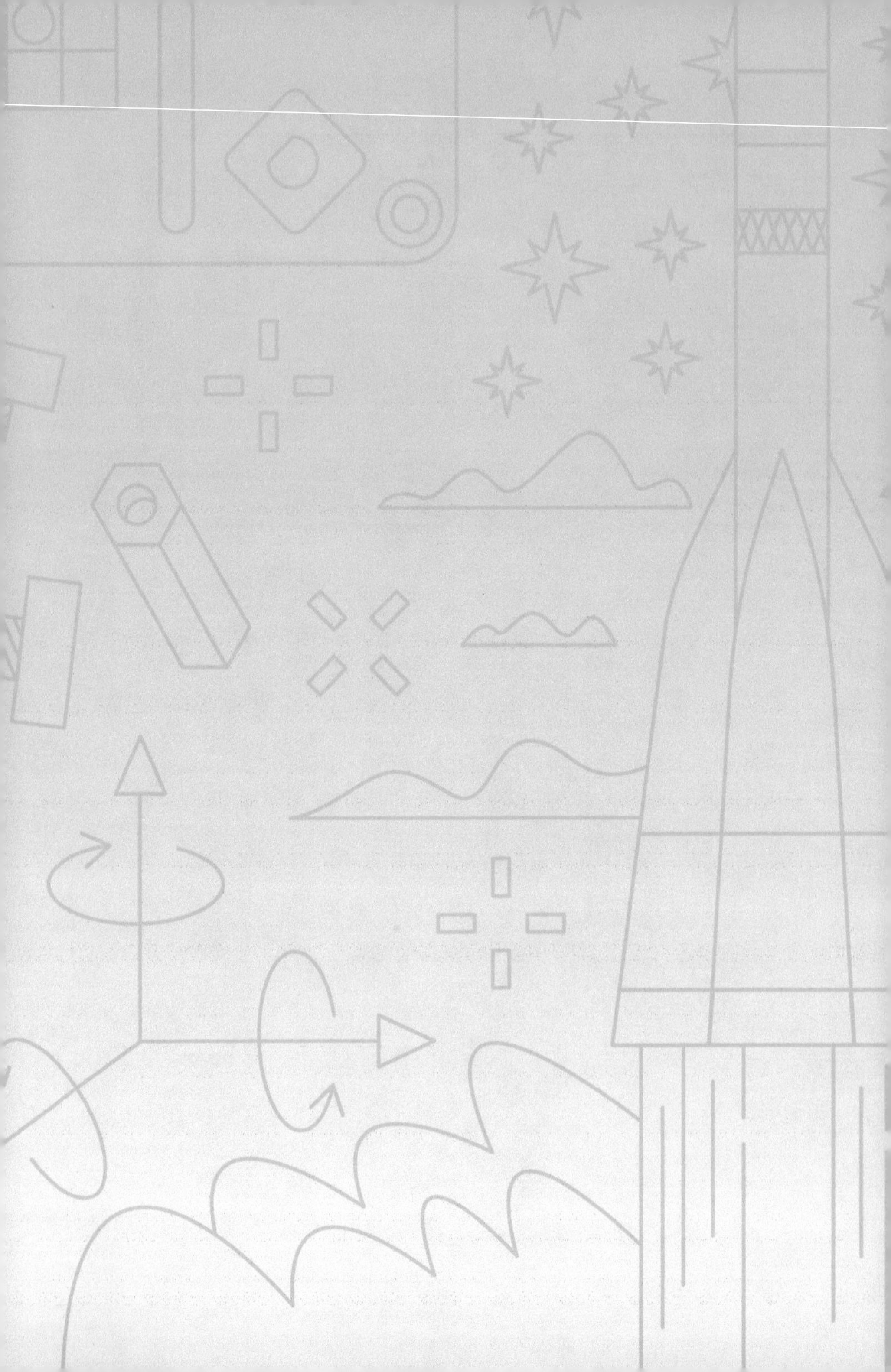

Chapter 6

Interactive pixel pet

Create a pint-sized pocket monster living in the digital world of a Raspberry Pi Sense HAT

Using sensors and output devices is a great way to make your computer programs more interactive. The Raspberry Pi Sense HAT contains a whole set of sensors that can detect movement, and in this activity, we'll use them to take a digital pet for a walk.

You'll need to design your pet avatar before you program any actions. There are examples of some famous characters you can make on an excellent sprite sheet created by Johan Vinet, which can be found at **rpimag.co/spritesheet**. You can use this to draw your space pet avatar, as seen in **Figure 6-1**.

Draw your picture out on squared paper using coloured pencils. You'll need two pet designs, the second preferably being very similar to the first so that we can animate your pet. In **Figure 6-2**, you can see that our two images are almost identical to each other — only the feet are in a different place.

Later, when you code your animation, this will create the illusion that the pet is walking.

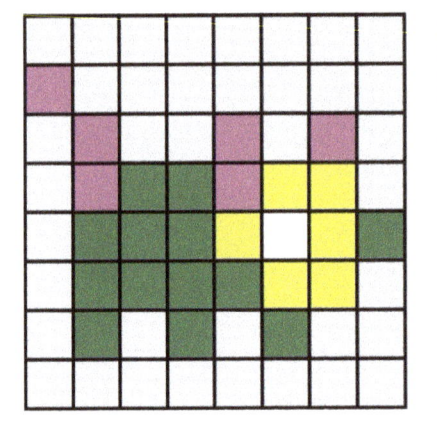

Figure 6-1
The first animation frame

Figure 6-2
The second animation frame

Labelling each pixel

Think of a letter from the alphabet to represent each colour in your pixel pet image: for example, **w** for white or **r** for red. If using squared paper for your design, you can write the letters on top, as shown in **Figure 6-3**. Note that **e** stands for empty.

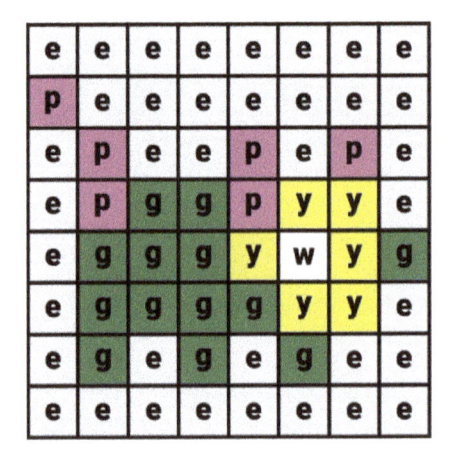

Figure 6-3 The pixel pet with colour labels

52 · Chapter 6 · Interactive pixel pet

You'll notice that we have eight rows and eight columns of letters, each separated by a comma, to make up the LED matrix on the Sense HAT. Repeat this step for your second pet design so that you end up with two grids of letters.

Code your pet

Now that you have your designs represented as letters in a grid, you can start to code them in Python. Open the Thonny editor (click the Raspberry Pi menu and select **Programming > Thonny**); if Thonny is already open, press the **New** button to start a new program. Save the new, empty file as **space-pet.py**.

First, you'll need to import all the modules and libraries required for this project into your code and create a **SenseHat** object:

```
from sense_hat import SenseHat
import time
sense = SenseHat()
```

Note that capital letters, full stops, and commas are very important in Python. Your code might not work if you don't include these. Next, create a variable for each colour label in your pet design, like this:

```
p = (204, 0, 204)  # Pink
g = (0, 102, 102)  # Dark Green
w = (200, 200, 200)  # White
y = (204, 204, 0)  # Yellow
e = (0, 0, 0)  # Empty
```

The numbers inside the brackets are the *RGB* values, or red, green, and blue values. Mixtures of these colours make different shades. The higher the number, the more of that colour it will contain. For example, **(255, 0, 0)** would make a solid red, whereas **(0, 255, 0)** would create a vivid green. You can change these numbers in your code to get the colours that you want.

Next, use a *list* to store your pixel pet design, like this:

```
pet1 = [
e, e, e, e, e, e, e, e,
p, e, e, e, e, e, e, e,
e, p, e, e, p, e, p, e,
e, p, g, g, p, w, w, e,
e, g, g, g, w, y, w, y,
e, g, g, g, g, w, w, e,
e, g, e, g, e, g, e, e,
e, e, e, e, e, e, e, e
]
```

Here you have created a variable called **pet1** and stored a list of labels for each colour by using **[** at the start and **]** at the end. Repeat for the second pixel pet design, using a different variable name like **pet2**. Your code should start to look something like the drawing you created earlier.

If you ran your code now, nothing would happen, because so far you have only told the program to store information. To make something happen, you will need to write a command to call on that data and display your colours in the correct order on the Sense HAT's LED matrix. Type this command underneath your lists:

```
sense.set_pixels(pet1)
```

Save your code by clicking **Save**, then run your program to test it works by clicking **Run** or pressing **F5**. Note what happens. Why did only one of your pet designs appear? It's because you only called **pet1** in your command.

Add a delay using the **time.sleep** function, then call the second picture using the same command as before, like this:

```
time.sleep(0.5)
sense.set_pixels(pet2)
```

Save and run your code to see your pet.

Animate your pet

So far, your pixel pet only changes once. To animate it fully, you will need to switch repeatedly between the two pictures with a time delay. You could write the commands out over and over again, but it makes more sense to put them in a loop.

Move to the end of your program and locate the **sense.set_pixels(pet1)** line and the two lines that follow. Change them to look like this:

```
for i in range(10):
    sense.set_pixels(pet1)
    time.sleep(0.5)
    sense.set_pixels(pet2)
    time.sleep(0.5)
```

Don't forget to add the extra **time.sleep(0.5)** on the last line, and remember to indent the lines after **for i in range(10):** to place them inside the **for** loop. This **for** loop with the **range** function will repeat the indented code ten times and then stop.

Save and run your code to watch the animation. You will notice that after the animation has finished, you are left with the same image still displayed on the LED matrix. There is a useful function that you can use to clear the LEDs; add this line above your new loop to clear the LEDs when you first run your program:

sense.clear(0, 0, 0)

Create a walking function

A *function* is a piece of code that you can use over and over. As the goal is to trigger the walking animation later on, it makes sense for us to put the animation code into a function that can be called when an action has been sensed by the hardware.

To put your code into a function, you simply need to add the line **def walking():** above your **for** loop and indent the lines beneath, like this:

```
def walking():
    for i in range(10):
        sense.set_pixels(pet1)
        time.sleep(0.5)
        sense.set_pixels(pet2)
        time.sleep(0.5)
```

The use of **def** here means that you are *defining* a function you have named **walking**. Now you need to call the function. So, at the bottom of your code, type:

```
walking()
```

Shake for more

It's time to use the Sense HAT's movement sensors, particularly its accelerometer, to trigger the walking function and make the project more interactive. Between **sense.clear(0, 0, 0)** and the function call to **walking()**, type:

```
sense.set_pixels(pet1)
while True:
    x, y, z = sense.get_accelerometer_raw().values()
    if x >= 2 or y >= 2 or z >= 2:
        break
```

The first line shows a (stationary) space pet, and the second creates an infinite loop. The next line obtains current movement readings from the Sense HAT on its x, y, and z coordinates. As your Raspberry Pi is presumably sitting still on a desk, those readings will have a very low value. The loop continues forever until the Sense HAT moves enough for the statement **x >= 2 and y >= 2 and z >= 2** to evaluate to **True**. When that happens, the **break** statement stops the loop. You can help the Sense HAT take an accelerometer reading above **2** by shaking it! Your final code is in the following listing and will look like **Figure 6-4**. Save your code and then run it. Nothing should happen until you shake your Raspberry Pi. Then the pet animation will run, and your pet will start to walk.

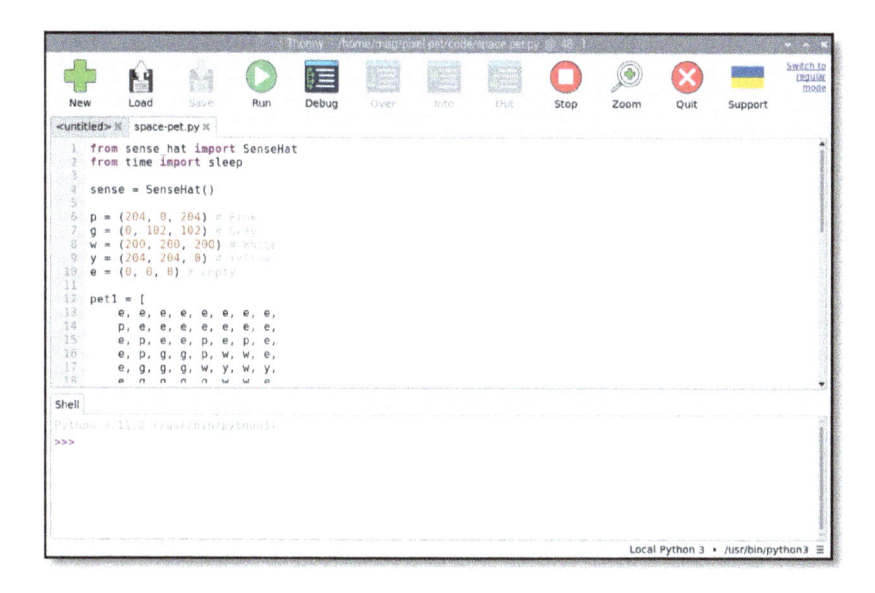

Figure 6-4 The final code running in Thonny

```
from sense_hat import SenseHat
import time

sense = SenseHat()

p = (204, 0, 204)    # Pink
g = (0, 102, 102)    # Dark Green
w = (200, 200, 200)  # White
y = (204, 204, 0)    # Yellow
e = (0, 0, 0)        # Empty
```

Experiment with the Sense HAT, 2nd Edition · 57

```
pet1 = [
e, e, e, e, e, e, e, e,
p, e, e, e, e, e, e, e,
e, p, e, e, p, e, p, e,
e, p, g, g, p, w, w, e,
e, g, g, g, w, y, w, y,
e, g, g, g, g, w, w, e,
e, g, e, g, e, g, e, e,
e, e, e, e, e, e, e, e
]
pet2 = [
e, e, e, e, e, e, e, e,
p, e, e, e, e, e, e, e,
e, p, e, e, p, e, p, e,
e, p, g, g, p, w, w, e,
e, g, g, g, w, y, w, y,
e, g, g, g, g, w, w, e,
e, e, g, e, g, e, e, e,
e, e, e, e, e, e, e, e
]

def walking():
    for i in range(10):
        sense.set_pixels(pet1)
        time.sleep(0.5)
        sense.set_pixels(pet2)
        time.sleep(0.5)

sense.clear(0, 0, 0)
sense.set_pixels(pet1)
while True:
    x, y, z = sense.get_accelerometer_raw().values()
    if x >= 2 or y >= 2 or z >= 2:
        break
walking()
```

Your pet will keep walking forever once you've shaken your Raspberry Pi for the first time. If you find your pet doesn't move, you can try changing the **if** statement to require less movement:

```
if x >= 1 or y >= 1 or z >= 1:
```

You could even add x, y, and z together to only take action when the sum of the raw values is over a certain amount:

```
if x + y + z >= 2:
```

Can you think of a way to modify the program so that the pet walks a bit after you shake it, and then stops walking until you shake it again?

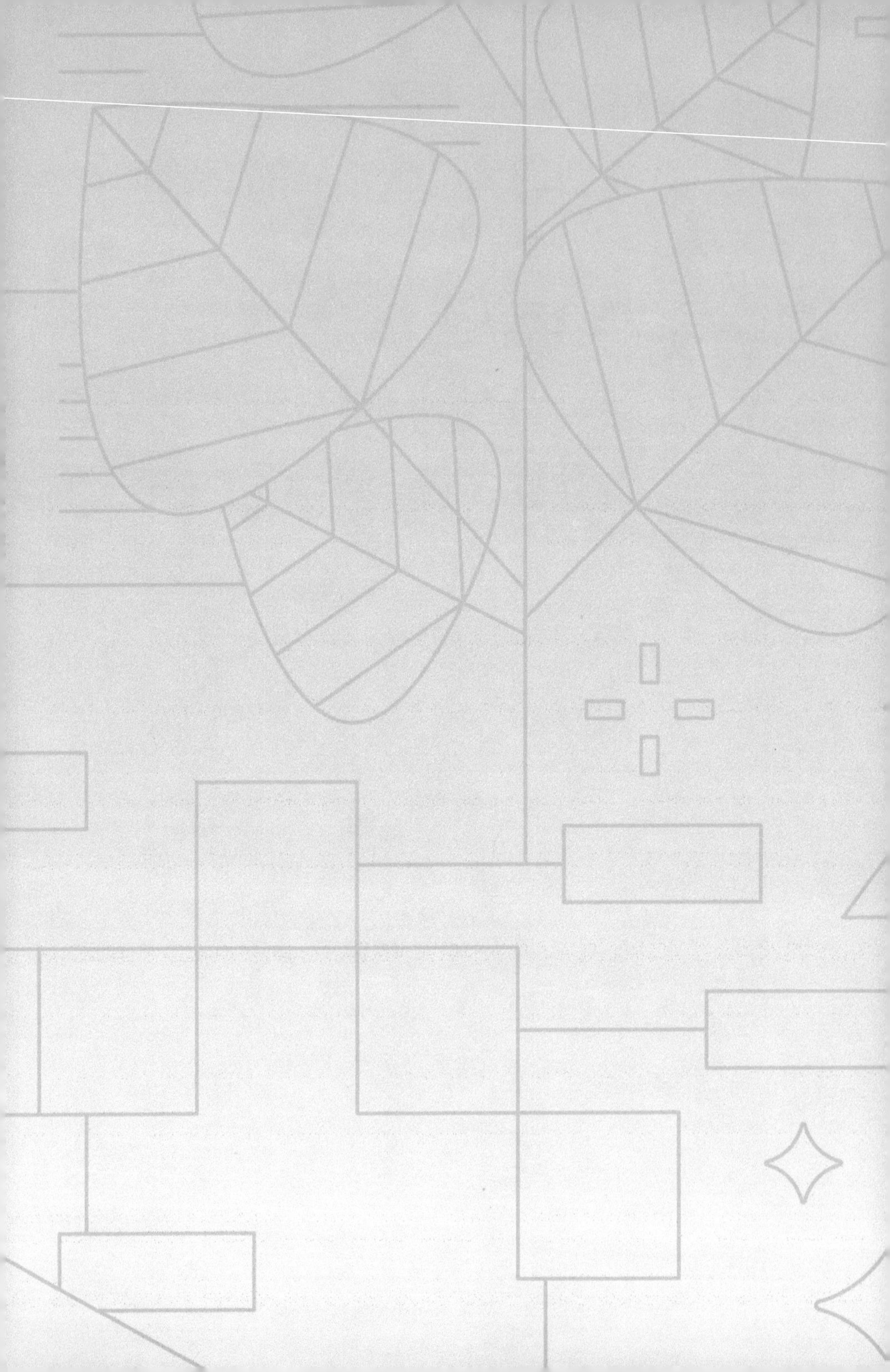

Chapter 7
Sense HAT data logger

Use your Sense HAT to take environmental readings, so you can walk around the house saying "sensors indicate…"

During the Astro Pi mission, a pair of Raspberry Pis with Sense HATs attached captured and logged a range of data about life on board the International Space Station.

In this activity, you will use a Raspberry Pi, a Sense HAT, and some Python code to create your own data-logging tool, which you can use to capture interesting data and perform experiments at home.

Using the sensors, we can measure the following conditions:

- Temperature
- Humidity
- Pressure
- Orientation
- Acceleration
- Magnetic field
- Colour and brightness (V2 Sense HAT only)

First, we'll write a short script to get data from the Sense HAT and output it to the screen. Open the Thonny editor (click the Raspberry Pi menu and select **Programming > Thonny**); if Thonny is already open, press the **New** button to start a new program. Save the new file as **datalogger.py**. To begin this script, you will need to import some Python modules to control your Sense HAT and fetch the date and time from the Raspberry Pi. Start by adding these three lines of code:

```
from sense_hat import SenseHat
from datetime import datetime
sense = SenseHat()
```

If you have a V2 Sense HAT, then you need to set up the colour sensor with the following extra lines of code.

```
sense.color.gain = 60
sense.color.integration_cycles = 64
```

Create a function that will fetch all the sensor data and return it as a list. Start by defining your function (**def get_sense_data():**) and creating an empty list (**sense_data = []**). The remaining function calls collect the data from the different sensors. In each case, you are appending the results to the **sense_datalist**.

```
def get_sense_data():
    sense_data = []
    # Get environmental data
    sense_data.append(sense.get_temperature())
    sense_data.append(sense.get_pressure())
    sense_data.append(sense.get_humidity())
    # Get colour sensor data (version 2 Sense HAT only)
    red, green, blue, clear = sense.colour.colour
    sense_data.append(red)
    sense_data.append(green)
    sense_data.append(blue)
    sense_data.append(clear)
    # Get orientation data
    orientation = sense.get_orientation()
    sense_data.append(orientation["yaw"])
    sense_data.append(orientation["pitch"])
```

```
    sense_data.append(orientation["roll"])
    # Get compass data
    mag = sense.get_compass_raw()
    sense_data.append(mag["x"])
    sense_data.append(mag["y"])
    sense_data.append(mag["z"])
    # Get accelerometer data
    acc = sense.get_accelerometer_raw()
    sense_data.append(acc["x"])
    sense_data.append(acc["y"])
    sense_data.append(acc["z"])
    # Get gyroscope data
    gyro = sense.get_gyroscope_raw()
    sense_data.append(gyro["x"])
    sense_data.append(gyro["y"])
    sense_data.append(gyro["z"])
```

Complete your data capture by adding the current date and time to the list.

```
    # Get the date and time
    sense_data.append(datetime.now())
```

The function should return the **sense_data** list at the end.

```
    return sense_data
```

To finish off, you can look at the data by printing out the list within an infinite loop. Add the following to the end of your script, then save and run the code.

```
while True:
    print(get_sense_data())
```

You should see a continuous stream of data in the shell, with each line looking something like this:

[38.94975280761719, 1012.20166015625, 32.560699462890625, 17, 10, 11, 20, 144.1204202422324, 12.432924190508864, 355.4826729655122, -33.784427642822266, -23.137866973876953, 3.40659761428833, -0.19746851921081543, -0.054034601897001266,

```
0.9491991996765137, -0.014246762730181217,
0.009623102843761444, -0.010858062654733658,
datetime.datetime(2022, 7, 25, 14, 26, 51, 341250)]
```

If you have an original Sense HAT without the colour sensor, you'll get an error unless you remove these lines:

```
# Get colour sensor data (version 2 Sense HAT only)
red, green, blue, clear = sense.colour.colour
sense_data.append(red)
sense_data.append(green)
sense_data.append(blue)
sense_data.append(clear)
```

Writing the data to a file

The program you have produced so far is able to continually check the Sense HAT sensors and write this data to the screen. We can make this more practical by altering the program to write the data to a *comma-separated value* (*CSV*) file instead, which you can examine once your logging program has finished. To create this file, you will need to do the following:

1. Create the file
2. Add a header row for each sensor reading
3. Write the data to the file

Import the **csv writer** class by adding the line **from csv import writer** to the top of the program:

```
from sense_hat import SenseHat
from datetime import datetime
from csv import writer
sense = SenseHat()
```

Remove your current **while True** loop. Replace it with these lines to open and write to a new .csv file:

```
with open('data.csv', 'w', buffering=1, newline='') as f:
    data_writer = writer(f)
```

Create a variable to hold the data from the function call within a new **while True** loop, then write that data to the new file. Make sure to set the file buffering mode to line mode (expressed as a **1**) so that every row is saved immediately and data loss is avoided.

```
while True:
    data = get_sense_data()
    data_writer.writerow(data)
```

Run your program for a few minutes. When you stop the program, you should be able to find the **.csv** file in your file manager.

The first line should look something like this:

```
36.324222564697266,1024.387939453125,32.6043815612793,0,0,0,1,
138.03485829553443,12.164214303276808,353.07380995988177,
-10.638025283813477,-9.077208518981934,1.978834867477417,
-0.20660144090652466,-0.11602965742349625,0.9455438256263733,
-0.005754658952355385,-0.00629773736000061,0.00323345884680748,
2022-07-26 11:12:45.316169
```

Adding a header to the CSV file

Your program is collecting so many different types of data in the CSV file that it can be hard to identify which type of data each column contains. To solve this problem, you can write a row to the top of the CSV file before you start the infinite loop.

Add a call to **data_writer.writerow** after you've created your **writer** object and before the **while True** loop starts. Pass it an array with the names of the columns you'd like to use. Your code (everything after **return sense_data**, that is) should look like this:

```
with open('data.csv', 'w', buffering=1, newline='') as f:
    data_writer = writer(f)
```

```
data_writer.writerow(['temp', 'pres', 'hum',
                      'r', 'g', 'b', 'clear', # V2 only
                      'yaw', 'pitch', 'roll',
                      'mag_x', 'mag_y', 'mag_z',
                      'acc_x', 'acc_y', 'acc_z',
                      'gyro_x', 'gyro_y', 'gyro_z',
                      'datetime'])
while True:
    data = get_sense_data()
    data_writer.writerow(data)
```

> **Tip**
>
> Make sure the headers are in the same order as the data produced by your `get_sense_data()` function. If you have an original Sense HAT without a colour sensor, be sure to remove the line that starts with `'r'`, `'g'`, `'b'`, `'clear'`.

Recording at specific time intervals

At the moment, your script records data as quickly as it possibly can. This is very useful for some experiments, but you may prefer to record data once per second, or even less frequently.

Normally, you would use a `sleep()` function to pause the script, but this can cause inaccurate readings from some of the IMU sensors. Instead, you can use `timedelta` to check the time difference between two readings.

At the top of your program, set the length of delay between readings in seconds, and fetch the current date and time.

```
timestamp = datetime.now()
delay = 1
```

Within your `while` loop, you can calculate the difference between the current time and the time stored in the `data` list (the last element, which we can refer to with the array index `-1`):

```
while True:
    data = get_sense_data()
    time_difference = data[-1] - timestamp
```

If the **time_difference** is greater than the delay you have set, the **data** can be written to the file. Change your **while** loop to look like this:

```
while True:
    data = get_sense_data()
    time_difference = data[-1] - timestamp
    if time_difference.seconds > delay:
        data_writer.writerow(data)
        timestamp = datetime.now()
```

While your program is running, open the terminal, use **cd** to change directory to the directory that contains the data.csv file, and run the command **tail -f data.csv** to watch as each line is added to the .csv file:

temp,pres,hum,r,g,b,clear,yaw,pitch,roll,mag_x,mag_y,mag_z,
acc_x,acc_y,acc_z,gyro_x,gyro_y,gyro_z,datetime
36.108428955078125,1024.423095703125,32.416011810302734,49,38,
38,90,138.01520101110313,12.227523326693655,352.8891865315218,
-29.801549911499023,-25.660537719726562,5.958069324493408,
-0.20684826374053955,-0.11651210486888885,0.9470059275627136,
-0.002123238518834114,0.0003891065716743469,
-0.0002552233636379242,2022-07-26 11:27:05.983233
36.23430633544922,1024.432373046875,33.37968826293945,49,38,
38,90,137.72729487720875,12.181723493214136,352.9463897927074,
-29.705188751220703,-25.5445613861084,6.508992671966553,
-0.20660144090652466,-0.11795946210622787,0.9484680891036987,
0.0003636479377746582,0.0006903782486915588,
-3.945082426071167e-06,2022-07-26 11:27:08.091969

If you have LibreOffice installed, you can use LibreCalc to open it as a spreadsheet. If you don't have LibreOffice installed, click the Raspberry Pi menu, choose **Preferences > Recommended Software**, and use the Recommended Software tool to install it. You should see that the file is only being written to periodically. You can adjust the delay time you have set to write more frequently or less frequently.

Starting your data logger on boot

This step is completely optional, but you might want to have your script run as soon as the Raspberry Pi boots up. To do this, you can use a **cron job**. Have a look at the section below to learn how to edit your **crontab** to start scripts on boot.

Sometimes you don't want to manually start a script that you have written. You may need the script to run once every hour, or maybe once every thirty seconds, or even every time your computer starts. On Linux-based systems like Raspberry Pi OS, this is a fairly easy task, as you can use a program called **cron**. Cron will run any command you tell it to run, whenever you have scheduled for it to do so. It will reference what is known as the *cron table*, which is normally abbreviated to *crontab*.

Editing the crontab

To open the crontab, you first need to open a terminal window. Then you can type:

```
crontab -e
```

The **-e** in this command is short for *edit*. If this is your first time opening your crontab, you'll be asked which text editor you would like to use:

```
rpi@raspberrypi:~ $ crontab -e
no crontab for rpi - using an empty one

Select an editor.  To change later, run 'select-editor'.
  1. /bin/nano        <---- easiest
  2. /usr/bin/vim.tiny
  3. /bin/ed

Choose 1-3 [1]:
```

Unless you have plenty of experience using **ed** or **vim**, the simplest editor to use is nano, so type **2** to select it and press **Enter**.

The nano editor is a simple command-line text editor. If you want to learn more about using nano, see **rpimag.co/nano**.

Syntax for cron

The crontab contains all the basic information you need to get started. Each line that starts with a **#** is a comment and is therefore ignored by the computer. At the bottom of the crontab, you should see a line that looks like this:

```
# m h  dom mon dow   command
```

- **m** is short for minute
- **h** is short for hour
- **dom** is short for day of the month
- **mon** is short for month
- **dow** is short for day of the week
- **command** is the shell command that you want to run

Creating a new cron job

To create a cron job, you need to decide under which circumstances you would like it to run. For instance, if you wanted to run a Python script on the 30th minute of every hour, you would write the following, making sure to change the **rpi** username to your own username and, if you saved **datalogger.py** in a different location, changing the path as needed:

```
30 * * * * python3 /home/rpi/my_script.py
```

If you wanted it to run every 30 minutes, you would use:

```
*/30 * * * * python3 /home/rpi/my_script.py
```

The **30** is telling the script to run every 30 minutes. The asterisks indicate that the script needs to run for all legal values for the other fields.

Here are a few more examples.

Run a script at 11:59 every Tuesday
```
59 11 * * 2 python3 /home/rpi/my_script.py
```

Run a script once a week on Monday
```
0 0 * * 1 python3 /home/rpi/my_script.py
```

Run a script at 12:00 on the 1st of January and June
```
0 12 1 1,6 * python3 /home/rpi/my_script.py
```

Run on boot

One incredibly useful feature of cron is the ability to run a command when the computer boots up. To do this, you use the **@reboot** syntax.

```
@reboot python3 /home/rpi/datalogger.py
```

Because the data logger runs continuously until you stop it, this is probably the best option. If you were to run it, for example, every 30 minutes, you'd have multiple data loggers trying to read the sensors and trying to write to the file, which would quickly become problematic!

If you want to quit your data logger after starting it from your crontab, you can run the following command from the terminal:

```
pkill -f datalogger.py
```

Edit and save the file

You can add your cron job to the bottom of the crontab. Save and exit nano by pressing **CTRL+X**, then **y** when you are prompted to save, and **ENTER** to accept the file name (don't change it).

Selecting the data to be recorded

You might not always want to record all the sensor data. One solution to this is to simply comment out the lines you don't need in your `get_sense_data()` function by adding hashtags (`#`) in front of them.

Can you set up your script to pass only the sensors you want to use into your `get_sense_data` function, ensuring that only data from those sensors is recorded? Don't forget to add a method to alter the header row of your CSV file as well!

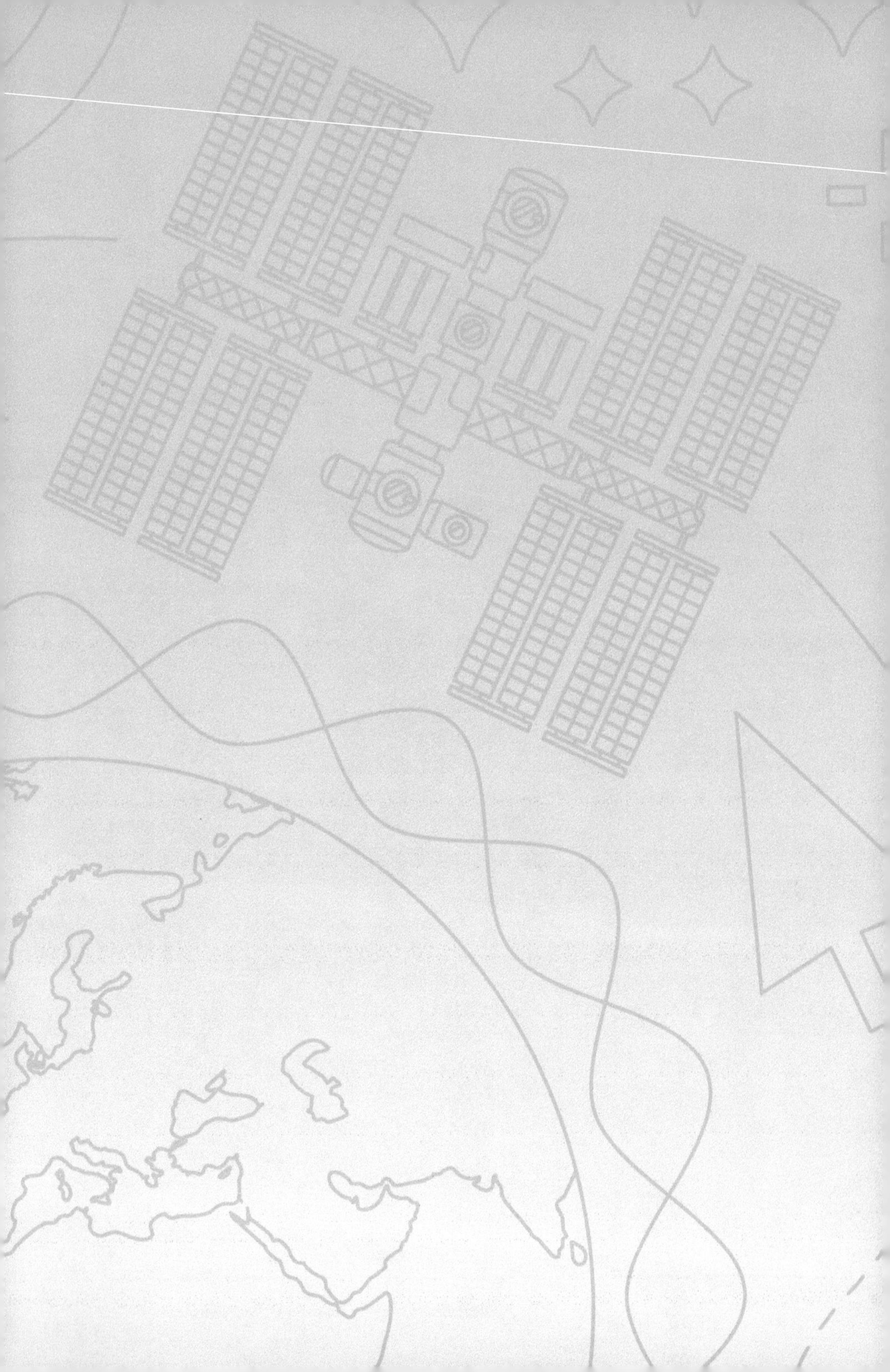

Chapter 8

Flappy Astronaut

Create your own clone of the Flappy Bird game using your Raspberry Pi, a Sense HAT, and some Python code

In this activity, you will create a game called Flappy Astronaut, which uses the Sense HAT joystick to navigate an astronaut through a series of pipes. Before you get to the core gameplay, you'll need to decide how to display your astronaut and the pipes on the Sense HAT's LED display. In Chapter 6, *Interactive pixel pet*, you used a list to design and display an interactive, animated space pet:

```
pet1 = [
e, e, e, e, e, e, e, e,
p, e, e, e, e, e, e, e,
e, p, e, e, p, e, p, e,
e, p, g, g, p, w, w, e,
e, g, g, g, w, y, w, y,
e, g, g, g, g, w, w, e,
e, g, e, g, e, g, e, e,
e, e, e, e, e, e, e, e
]
```

The only problem with using a single list like this is that it can be difficult to figure out which item in the list corresponds to which pixel on the screen. For instance, what is the list index of the pixel at **x = 5** and **y = 5**? It can be calculated, but it is a little tricky.

A better way to plot your pixels

To get around this problem, programmers often use two-dimensional lists, also known as *lists of lists*, to represent the arrangement of pixels on a screen. Here is a simple list of lists to describe a noughts and crosses (tic-tac-toe) board.

```
board = [['X', 'O', 'X'],
         ['O', 'X', 'O'],
         ['O', 'O', 'X']]
```

This is a great way to represent the board because you can easily use **x** and **y** coordinates to find out what is in each of the squares. To find out the character in any given position, you can just write **board[y][x]**.

For instance, if you want to find out which character is in the bottom-left corner, you already know that it has an **x** position of **0** (don't forget that in Python, we start counting items in a list from **0**) and a **y** position of **2**. So, in this example, that would be **board[2][0]**.

Why is **y** not **0**? The topmost row is the first (0th) row in the array, which means our y values increase as you go down, rather than increasing as you go up (like the coordinate system you probably learnt in school).

Using 2D lists with the Sense HAT

To store the colour information, you can use a pair of tuples: one for red and one for blue. Start a new program and add the following lines to it:

```
from sense_hat import SenseHat
sense = SenseHat()

RED = (255, 0, 0)
BLUE = (0, 0, 255)
```

Now you are going to make a list of lists filled with the variable **BLUE**. Manually creating it would require a lot of typing, but you can use a *list comprehension* (see "Python list comprehension" on page 76) instead to complete the task in a single line. Add this line to the end of your program:

```
matrix = [[BLUE for column in range(8)] for row in range(8)]
```

What does this code do? The section **[BLUE for column in range(8)]** creates one list with eight values of **(0, 0, 255)** inside it, and the **for row in range(8)** part makes eight copies of that list inside another list. You can run the command **print(matrix)** to see the result for yourself.

However, you can't use a list of lists with the Sense HAT, as its software only understands a flat, one-dimensional list. To deal with this issue, you are going to create a function that turns 2D lists into 1D lists. You can then use this function every time the matrix needs to be displayed.

To flatten a 2D list into a 1D list, you can again use a list comprehension. Here's how to flatten a list (don't add this to your program, though):

```
flattened = [pixel for row in matrix for pixel in row]
```

What does this do? The **for row in matrix** part looks at each of the lists in the matrix, and the **for pixel in row** section looks at the individual pixels in each row of that list. These pixels are then all placed into a single list.

You can turn this into a function to avoid having to write it out all the time. Add the following to your program:

```
def flatten(matrix):
    flattened = [pixel for row in matrix for pixel in row]
    return flattened
```

To flatten your matrix and then display it on the Sense HAT, you can now simply add these lines of code to the bottom of your file.

```
matrix = flatten(matrix)
sense.set_pixels(matrix)
```

Save and run the program, and you should see all your pixels turn blue!

Python list comprehension

If you want to generate a list using Python, it is quite easy to do so using a **for** loop:

```
new_list = []
for i in range(10):
    new_list.append(i)
print(new_list)
```

This results in the following output:

```
[0, 1, 2, 3, 4, 5, 6, 7, 8, 9]
```

The same list can be created in a single line using a construct that exists in many programming languages: a list comprehension:

```
new_list = [i for i in range(10)]
```

You can use any *iterable* in a list comprehension, so making a list from another list is easy:

```
numbers = [1, 2, 3, 4, 5]
numbers_copy = [number for number in numbers]
print(numbers_copy)
```

This displays:

```
[1, 2, 3, 4, 5]
```

You can also do calculations within a list comprehension:

```
numbers = [1, 2, 3, 4, 5]
double = [number * 2 for number in numbers]
print(double)
```

The results are:

```
[2, 4, 6, 8, 10]
```

You can implement string operations as well:

```
verbs = ['shout', 'walk', 'see']
present_participle = [word + 'ing' for word in verbs]
print(present_participle)
```

The output of that is:

```
['shouting', 'walking', 'seeing']
```

You can also extend lists quite easily:

```
animals = ['cat', 'dog', 'fish']
animals = animals + [animal.upper() for animal in animals]
print(animals)
```

This produces the following:

```
['cat', 'dog', 'fish', 'CAT', 'DOG', 'FISH']
```

Generate pipes

In Flappy Astronaut, the astronaut will have to avoid 'pipes' that sprout from the top and bottom of the matrix. The colour of the pipes is going to be red.

To begin, you can create a single column of red pixels on the right-hand side of the matrix, as shown in **Figure 8-1**.

All you need to do is set the last item in each list within the matrix to **RED** instead of **BLUE**.

You can use a **for** loop so that, for each list in the matrix, the last item is set to **RED**. Position this for loop so that it runs before you flatten and display the matrix. You can find the answer at the end of "Index a list in Python" on page 79 if you need it.

Figure 8-1   *A single column of pixels on the Sense HAT*

Here is an example of what your code should look like, with a comment showing you where your **for** loop should go.

```
from sense_hat import SenseHat

sense = SenseHat()
RED = (255, 0, 0)
BLUE = (0, 0, 255)

matrix = [[BLUE for column in range(8)] for row in range(8)]

def flatten(matrix):
    flattened = [pixel for row in matrix for pixel in row]
    return flattened

# Place your for loop here

matrix = flatten(matrix)
sense.set_pixels(matrix)
```

Index a list in Python

A Python list is a type of data structure. It can hold collections of any data type, and even a mixture of data types. Here is an example of a list of strings in Python:

```
band = ['paul', 'john', 'ringo', 'george']
```

In Python, lists are indexed from **0**. This means you can talk about the zeroth item in a list. In our example, the zeroth item is **'paul'**. To find the value of an item in a list, you simply type the name of the list, followed by the index:

```
print(band[0])
'paul'
print(band[2])
'ringo'
```

To find the value of the last item in a Python list, you can use the index **-1**:

```
print(band[3])
'george'
print(band[-1])
'george'
```

You can find the value of the penultimate item using the index **-2**, and so on.

Sometimes you might want to use a two-dimensional list, i.e. a list of lists. In that case, you will have to provide two indices to find a specific item. Here is a list representing a noughts and crosses game:

```
board = [['X', 'O', 'X'],
         ['O', 'X', 'O'],
         ['O', 'O', 'X']]
```

To find the central character in this list of lists, you would use **board[1][1]**. Here's how you'd set the last column in the matrix to **RED**:

```
for row in matrix:
    row[-1] = RED
```

Mind the gaps

Now that you have a column of pixels representing a pipe on the right-hand side of the matrix, you need to insert a gap into it through which the astronaut can fly. The gap needs to be three pixels high and should be placed randomly in the column of red pixels, as shown in **Figure 8-2**.

Figure 8-2 A gap in the red pixel column

You'll want the three-pixel-high gap to be centred around one of the rows between **1** and **6** (inclusive). You can use the **random** module to achieve this. The **random** module is one of the standard modules in Python; you can use it to create pseudo-random numbers in your code (see "Generate random numbers" on page 39 for more details).

Here's what you need to do:

1. Import the **randint** method at the top of your code
2. After the for loop has ended, create a variable called **gap**, and assign a random number between **1** and **6** as its value
3. Change the last pixel in that row of the matrix to **BLUE**
4. Change the last pixel in row **gap - 1** to **BLUE**
5. Change the last pixel in row **gap + 1** to **BLUE**

Run the following program to draw a pipe with a gap at a random location:

```python
from sense_hat import SenseHat
from random import randint

sense = SenseHat()
RED = (255, 0, 0)
BLUE = (0, 0, 255)

matrix = [[BLUE for column in range(8)] for row in range(8)]

def flatten(matrix):
    flattened = [pixel for row in matrix for pixel in row]
    return flattened

for row in matrix:
    row[-1] = RED
gap = randint(1, 6)
matrix[gap][-1] = BLUE
matrix[gap + 1][-1] = BLUE
matrix[gap - 1][-1] = BLUE

matrix = flatten(matrix)
sense.set_pixels(matrix)
```

Create a function to make more pipes

The game would be too easy if only one set of pipes was created. You can generate as many pipes as you like by using a function.

Below your **flatten()** function, create a new function called **gen_pipes**:

`def gen_pipes(matrix):`

Put the code you wrote to generate the first set of pipes into this function; you can just add some indentation to do this. At the end of the function, you should **return** the altered **matrix**:

```
def gen_pipes(matrix):
    for row in matrix:
        row[-1] = RED
    gap = randint(1, 6)
    matrix[gap][-1] = BLUE
    matrix[gap + 1][-1] = BLUE
    matrix[gap - 1][-1] = BLUE
    return matrix
```

Then, call the function before you flatten and display the **matrix**:

```
matrix = gen_pipes(matrix)
matrix = flatten(matrix)
sense.set_pixels(matrix)
```

Here's what your code should look like now:

```
from sense_hat import SenseHat
from random import randint

sense = SenseHat()
RED = (255, 0, 0)
BLUE = (0, 0, 255)

matrix = [[BLUE for column in range(8)] for row in range(8)]

def flatten(matrix):
    flattened = [pixel for row in matrix for pixel in row]
    return flattened

def gen_pipes(matrix):
    for row in matrix:
        row[-1] = RED
    gap = randint(1, 6)
    matrix[gap][-1] = BLUE
    matrix[gap + 1][-1] = BLUE
    matrix[gap - 1][-1] = BLUE
    return matrix

matrix = gen_pipes(matrix)
```

```
matrix = flatten(matrix)
sense.set_pixels(matrix)
```

Moving pipes algorithm

Now that you can generate as many pipes as you want, you'll need to move them across the matrix so that they proceed towards the left of the screen as shown in **Figure 8-3**.

Figure 8-3   The pipe moving across the LED display

It might be easier to picture this on a smaller scale, like on a 5×5 matrix:

```
  0 1 2 3 4
0 b b b b r
1 b b b b r
2 b b b b b
3 b b b b r
4 b b b b r
```

To move the red pixels (**r**) to the left, you can follow a simple algorithm, expressed in pseudocode as:

```
Move all the items at index 1 in each of the rows to index 0
Move all the items at index 2 in each of the rows to index 1
Move all the items at index 3 in each of the rows to index 2
Move all the items at index 4 in each of the rows to index 3
Set all the items at index 5 in each row to b
```

This would then give you a matrix that looks like this:

```
  0 1 2 3 4
0 b b b r b
1 b b b r b
2 b b b b b
3 b b b r b
4 b b b r b
```

You could run the algorithm again to repeat the movement, which would return the following:

```
  0 1 2 3 4
0 b b r b b
1 b b r b b
2 b b b b b
3 b b r b b
4 b b r b b
```

If you do this with your matrix, the red line will start to move from right to left.

Move the pipes

As mentioned in the previous section, the algorithm to move the pipes will need to be repeated each time you want to shift the pixels left by **1**. Any code that needs to be repeated can be placed inside a function. Create this function below your **gen_pipes(matrix)** function:

```
def move_pipes(matrix):
```

The algorithm for this function can be broken down into the following pseudocode:

```
for each row in the matrix:
    for each item in the row from 0 to 7
        set the item to be the same as the next item in the row
    set the last item in the row to BLUE
```

Within the function, you can implement this in a **for** loop, like this:

```
def move_pipes(matrix):
    for row in matrix:
        for i in range(7):
            row[i] = row[i + 1]
        row[-1] = BLUE
    return matrix
```

Watch the pipes move

At this stage, running your code won't do much; you'll need to call your functions in a loop to see it working. Right now, you should have these three lines at the bottom of your code:

```
matrix = gen_pipes(matrix)
matrix = flatten(matrix)
sense.set_pixels(matrix)
```

Rather than using two lines of code to flatten the matrix and then display it, you can just use a single line. This will avoid flattening the actual matrix each time by creating a flattened version of the matrix for the display instead. Replace the last three lines with this:

```
matrix = gen_pipes(matrix)
sense.set_pixels(flatten(matrix))
```

Now you can add your **move_pipes(matrix)** function call:

```
matrix = gen_pipes(matrix)
sense.set_pixels(flatten(matrix))
matrix = move_pipes(matrix)
```

Although this will move the pipes, they won't yet be displayed, as there is no second **set_pixels** call. To solve this, you can add in a loop so that moving and displaying always follow each other. But first, you'll need to insert a delay to keep it from moving too fast. At the top of your code, import the **sleep** method from the **time** module:

```
from time import sleep
```

Next, put the last two lines in a loop and add a **sleep** command at the end:

```
matrix = gen_pipes(matrix)
for i in range(9):
    sense.set_pixels(flatten(matrix))
    matrix = move_pipes(matrix)
    sleep(1)
```

One small alteration will give you a continuous stream of pipes. Simply change the **for** loop to repeat four times, then enclose the entire last section of code in an infinite **while True** loop. This will call **gen_pipes()** after you've moved the playfield four times, which draws a new pipe into the rightmost column:

```
while True:
    matrix = gen_pipes(matrix)
    for i in range(4):
        sense.set_pixels(flatten(matrix))
        matrix = move_pipes(matrix)
        sleep(1)
```

Run the program now, and you should see an endlessly scrolling playfield with random pipes every four pixels. It will keep running until you stop the program, as shown in **Figure 8-4**.

Add your astronaut

Your astronaut will be represented by a single coloured pixel. You can choose any colour you like for the astronaut, but the example will use yel-

Figure 8-4 Multiple pipes moving across the LED display

low. Where you have set your other colour variables, create a new tuple for your chosen astronaut colour:

```
RED = (255, 0, 0)
BLUE = (0, 0, 255)
YELLOW = (255, 255, 0)
```

As the astronaut is going to be a single pixel, they'll need an **x** and a **y** coordinate so that the pixel at those coordinates can be illuminated. Near where you have set your colours, set an **x** and **y** position for the astronaut:

```
x = 0
y = 0
```

The player is going to control the astronaut using the Sense HAT's joystick. The joystick can be set up so that whenever it is moved, it sends the event to a function you have created. Events can be things such as 'pressed up' or 'released right' (see "Trigger function calls with the Sense HAT joystick" on page 88). Just above your **while True** loop, add in some code for the joystick to use a function (one which you have not yet created):

```
sense.stick.direction_any = move_astronaut
```

Now you need to create a **move_astronaut** function. This will have a single parameter, which is the event. Create the function below one of your other functions.

This function is going to need to alter the **x** and **y** variables you set for the astronaut's position. In Python, a function is not normally allowed to alter the value of variables that have been declared outside of itself. To allow your **move_astronaut** function to set the **x** and **y** variables, you need to state that **x** and **y** are global variables.

You'll need to hide the astronaut before you draw them again, then check for a **pressed** event and change **x** or **y** based on which direction you moved the joystick in. Note that the 'if' statements also check to make sure that the movement wouldn't place the astronaut off the edge of the display — this is to prevent your code from crashing due to an error:

```
def move_astronaut(event):
    global x, y
    sense.set_pixel(x, y, BLUE) # Hide the astronaut
    if event.action == "pressed":
        if event.direction == "up" and y > 0:
            y -= 1
        elif event.direction == "down" and y < 7:
            y += 1
        elif event.direction == "right" and x < 7:
            x += 1
        elif event.direction == "left" and x > 0:
            x -= 1
    sense.set_pixel(x, y, YELLOW) # Show the astronaut
```

You will also need to draw the astronaut each time the playfield scrolls. Add the following line near the end of your loop, just before **sleep(1)**:

```
sense.set_pixel(x, y, YELLOW) # Show the astronaut
```

Trigger function calls with the Sense HAT joystick

The Sense HAT joystick can be used to trigger function calls in response to being moved.

For instance, you can tell your program to continually 'listen' for a specific event, such as the joystick being pushed up (**direction_up**), and to then trigger a function (called **pushed_up** in this example) in response. The function triggered by the event can either have no parameters or take the event as a parameter. In the example below, the event is simply printed out.

```
def pushed_up(event):
    print(event)

sense.stick.direction_up = pushed_up
```

This function would print a timestamp of the event, the direction in which the joystick was moved, and the specific action. The output would look like this:

```
InputEvent(timestamp=1503565327.399252, direction=u'up',
action=u'pressed')
```

Another useful example is the **direction_any** method: if you use this method, as seen in the example below, the **do_thing** function will be triggered in response to any joystick event. For instance, you could define the **do_thing** function so that it reports the exact event in plain English.

```
def do_thing(event):
    if event.action == 'pressed':
        print('You pressed me')
        if event.direction == 'up':
            print('Up')
        elif event.direction == 'down':
            print('Down')
    elif event.action == 'released':
        print('You released me')

sense.stick.direction_any = do_thing
```

Collisions, or GAME OVER

To finish off the game, you need to ensure that it ends whenever the astronaut collides with one of the pipes. All you need to do is check whether the astronaut's x, y position corresponds to a red item in the matrix. If so, you'll need to end the game!

First, create a new variable called **game_over**, and set it to **False** near where you have set your colour constants. This variable will indicate whether the game has ended:

```
game_over = False
```

Next, add the following new function to your program. You can call this function to find out when the astronaut has collided with a pipe:

```
def check_collision():
    global game_over
    if matrix[y][x] == RED:
        game_over = True
```

You should call it at the end of the **move_astronaut** function, so add this line near the end, just after the call to **sense.set_pixel(x, y, YELLOW)**. If you put it before that call, the pixel you're checking would never be red!

```
    check_collision()
```

You're almost there! Now you need to check collisions from inside your loop and exit your loop when **game_over** becomes 'True'. Because you need to exit while you're two loops deep (i.e. within both a **while** and a **for** loop), the most expedient way to exit is to raise a **SystemExit** exception, which tells Python to stop running.

But before you do that, you should let the player know what happened! Add the following lines to your **for** loop, just after the call to **sense.set_pixels(flatten(matrix))**:

```
        check_collision()
        if game_over:
```

```
        sense.show_message('Game Over')
        raise SystemExit
```

When your astronaut collides with the pipes, this message should scroll, and your program will end. This means you've officially finished the program and should now have a working game you can play!

Here's the final listing for Flappy Astronaut:

```
from sense_hat import SenseHat
from random import randint
from time import sleep

sense = SenseHat()
RED = (255, 0, 0)
BLUE = (0, 0, 255)
YELLOW = (255, 255, 0)
game_over = False

x = 0
y = 0

matrix = [[BLUE for column in range(8)] for row in range(8)]

def flatten(matrix):
    flattened = [pixel for row in matrix for pixel in row]
    return flattened

def gen_pipes(matrix):
    for row in matrix:
        row[-1] = RED
    gap = randint(1, 6)
    matrix[gap][-1] = BLUE
    matrix[gap + 1][-1] = BLUE
    matrix[gap - 1][-1] = BLUE
    return matrix

def move_pipes(matrix):
    for row in matrix:
        for i in range(7):
```

```
            row[i] = row[i + 1]
        row[-1] = BLUE
    return matrix

def move_astronaut(event):
    global x, y
    sense.set_pixel(x, y, BLUE) # Hide the astronaut
    if event.action == "pressed":
        if event.direction == "up" and y > 0:
            y -= 1
        elif event.direction == "down" and y < 7:
            y += 1
        elif event.direction == "right" and x < 7:
            x += 1
        elif event.direction == "left" and x > 0:
            x -= 1
    check_collision()
    sense.set_pixel(x, y, YELLOW) # Show the astronaut

def check_collision():
    global game_over
    if matrix[y][x] == RED:
        game_over = True

sense.stick.direction_any = move_astronaut

while True:
    matrix = gen_pipes(matrix)
    for i in range(4):
        sense.set_pixels(flatten(matrix))
        check_collision()
        if game_over:
            sense.show_message('Game Over')
            raise SystemExit
        matrix = move_pipes(matrix)
        sense.set_pixel(x, y, YELLOW) # Show the astronaut
        sleep(1)
```

Challenge: Taking it further

Here are some ideas for how you could improve your game:

The for loop alters the spacing between the pipes — try changing its range to see if that makes the game trickier. Want to make it even harder? Shorten the amount of time your program sleeps for to make the pipes move faster.

Try altering the game to be a little more like Flappy Bird: can you make it so that the astronaut is falling until the flick of the joystick causes them to move upwards? You could also add a scoring system so that each time you go through the for loop, the score increases by 1, displaying the final number once the game ends.

Navigate your astronaut past a pipe, then move them backwards to crash into the one they just cleared. Does the game end? Can you figure out why it might fail to detect the collision and correct it?